Introduction to Vector Analysis

Seventh Edition

Harry F. Davis
Arthur David Snider

Student Solutions Manual

Hawkes Learning Systems

Copyright ©2008 by Hawkes Learning Systems/Quant Systems, Inc. All rights reserved.
1023 Wappoo Rd. A-6, Charleston, SC 29407
www.hawkeslearning.com

All rights reserved. No part of this publication may be reproduced, stored in a retrieval system, or transmitted in any form or by any means, electronic, mechanical, photocopying, recording, or otherwise, without the prior written permission of the publisher.

Printed in the United States

ISBN 13: 978-1-932628-31-9
ISBN 10: 1-932628-31-2

Table of Contents

Chapter 1: Vector Algebra 1

Chapter 2: Vector Functions of a Single Variable 23

Chapter 3: Scalar and Vector Fields 35

Chapter 4: Line, Surface, and Volume Integrals 53

Chapter 5: Advanced Topics 75

Appendices 93

Chapter 1

Vector Algebra

1.1 Definitions

No Problems.

1.2 Addition and Subtraction

The first four problems refer to figure 1.6.

1. Write **C** in terms of **E**, **D**, **F**.
   Solution: Because $\mathbf{C} + \mathbf{D} - \mathbf{E} + \mathbf{F} = 0$, $\mathbf{C} = -\mathbf{D} + \mathbf{E} - \mathbf{F}$

3. Solve for **x** : $\mathbf{x} + \mathbf{B} = \mathbf{F}$
   Solution: There are several possible solutions, and the simplest is $\mathbf{x} = \mathbf{F} - \mathbf{B} = \mathbf{A}$. Because **K** is equal to and in the opposite direction of **A**, we can also say $\mathbf{x} = \mathbf{F} - \mathbf{B} = -\mathbf{K}$.

5. If **A** and **B** are represented by arrows whose initial points coincide, what arrow represents **A**+**B**?
   Solution: An arrow that connects the initial point of **A** to the terminal point of **B** after the initial point of **B** has been moved to the terminal point of **A**.

7. Is the following statement correct? If **A**, **B**, **C** and **D** are distinct nonzero vectors represented by arrows from the origin to the points A, B, C and D respectively, and if $\mathbf{B} - \mathbf{A} = \mathbf{C} - \mathbf{D}$, then $ABCD$ is a parallelogram.
   Solution: Yes, because if $\mathbf{B} - \mathbf{A} = \mathbf{C} - \mathbf{D}$ these are parallel vectors of equal magnitude, so $ABCD$ is a parallelogram.

1.3 Multiplication of Vectors by Numbers

1. Is it ever possible to have $|\mathbf{A}| < 0$?
   Solution: No, length is always a positive number.

1.4. CARTESIAN COORDINATES

3. If $\mathbf{A}$ is a nonzero vector, and if $s = |\mathbf{A}|^{-1}$, what is $|-s\mathbf{A}|$?
   Solution: $s = 1/|\mathbf{A}|$ so $|-s\mathbf{A}| = |s\mathbf{A}| = \left|\frac{\mathbf{A}}{|\mathbf{A}|}\right|$ is the magnitude of a unit vector, 1.

5. If $\mathbf{A}$ is a scalar multiple of $\mathbf{B}$, is $\mathbf{B}$ necessarily a scalar multiple of $\mathbf{A}$?
   Solution: No, because for the zero vector this is not necessarily so. However, for nonzero vectors, if $\mathbf{A} = s\mathbf{B}$ then $(1/s)\mathbf{A} = \mathbf{B} = t\mathbf{A} = \mathbf{B}$ for $s \neq 0$, proving $\mathbf{B}$ is a scalar multiple of $\mathbf{A}$.

7. If $|\mathbf{A}| = |\mathbf{B}|$, is it necessarily true that $\mathbf{A} = \mathbf{B}$?
   Solution: No, vectors have two components, a direction and a length. If two vectors have the same length it is not necessary for them to point in the same direction.

9. How many distinct vectors exist, all having unit magnitude, perpendicular to a given line in space?
   Solution: In two dimensions there are only two free unit vectors perpendicular to a given line, and they point in opposite directions. In three dimensional space, there are an infinite number of unit vectors perpendicular to any line.

11. Let $\mathbf{A}$ and $\mathbf{B}$ be nonzero vectors represented by arrows with the same initial point to points A and B respectively. Let $\mathbf{C}$ denote the vector represented by an arrow from this same initial point to the midpoint of the line segment AB. Write $\mathbf{C}$ in terms of $\mathbf{A}$ and $\mathbf{B}$.
    Solution: $\mathbf{C} = \frac{\mathbf{A}}{2} + \frac{\mathbf{B}}{2}$

13. Find nonzero scalars a, b, and c such that $a\mathbf{A} + b(\mathbf{A} - \mathbf{B}) + c(\mathbf{A} + \mathbf{B}) = 0$ for every pair of vectors $\mathbf{A}$ and $\mathbf{B}$.
    Solution: Multiply the coefficients out to get $a\mathbf{A} + b\mathbf{A} - b\mathbf{B} + c\mathbf{A} + c\mathbf{B} = 0$, then group the coefficients of the vectors to get $(a + b + c)\mathbf{A} + (c - b)\mathbf{B} = 0$. Now solve the simultaneous system of equations $\begin{array}{l} a+b+c = 0 \\ b-c = 0 \end{array}$ to find the relations among the coefficients. We find $b = c$ and $a = -2b$, so in particular we can choose any value for any one of the variables and compute the others. Choosing $a = 2$, for example, gives $a = 2$, $b = -1$, $c = -1$.

1.4 Cartesian Coordinates

1. What is the x component of $\mathbf{i}$?
   Solution: "The x component" means the expression multiplying the unit vector $\mathbf{i}$, so the x component of $\mathbf{i}$ is 1.

3. What is the magnitude of $\mathbf{i} + \mathbf{j}$?
   Solution: The magnitude of a vector is the square root of the sum of the squares of its x, y, and z components. In this case, $|\mathbf{i} + \mathbf{j}| = \sqrt{1^2 + 1^2} = \sqrt{2}$.

5. With the axes in conventional position (fig. 1.9), directions may be specified in geographical terms. What is the unit vector pointing west? south? northeast?
   Solution: The unit vectors in these directions are $-\mathbf{i}$, $-\mathbf{j}$, and $\frac{\sqrt{2}}{2}\mathbf{i} + \frac{\sqrt{2}}{2}\mathbf{j}$, respectively.

7. The direction of a nonzero vector in the plane can be described by giving the angle θ it makes with the positive x direction (see fig. 1.11). This angle is conventionally taken to be positive in the counterclockwise sense. Write A_1 and A_2 in terms of $|\mathbf{A}|$ and this angle θ.
   Solution: $A_1 = |\mathbf{A}|\cos\theta$, and $A_2 = |\mathbf{A}|\sin\theta$, where $\mathbf{A} = A_1\mathbf{i} + A_2\mathbf{j}$.

9. In terms of $\mathbf{i}$ and $\mathbf{j}$, determine

    (a) the unit vector at positive angle $60°$ with the x axis.
    (b) the unit vector with $\theta = -30°$ (θ as in exercise 7).
    (c) the unit vector having the same direction as $3\mathbf{i} + 4\mathbf{j}$.
    (d) the unit vectors having x components equal to $1/2$.
    (e) the unit vectors perpendicular to the line $x + y = 0$.

    Solution:

(a) Take the cosine and sine of 60°, multiply these by the unit vectors in the x and y direction and add to get $\frac{1}{2}\mathbf{i} + \frac{\sqrt{3}}{2}\mathbf{j}$

(b) Again, we are finding a unit vector with a particular angle relative to the x axis, so multiply $\mathbf{i}$ by the cosine of 30° and $\mathbf{j}$ by the sine of 30° and add these to get $\frac{\sqrt{3}}{2}\mathbf{i} - \frac{1}{2}\mathbf{j}$

(c) We are given a vector in the proper direction. All we need to do is divide it by its length to create a unit vector. The length of $3\mathbf{i} + 4\mathbf{j}$ is $\sqrt{3^2 + 4^2} = \sqrt{25} = 5$, so the unit vector in the direction of $3\mathbf{i} + 4\mathbf{j}$ is $\frac{3}{5}\mathbf{i} + \frac{4}{5}\mathbf{j}$

(d) Because we know $x = 1/2$, the vector must look like $\frac{1}{2}\mathbf{i} + b\mathbf{j}$. The length of a unit vector must be 1, so we solve $\sqrt{(1/2)^2 + b^2} = 1$ to find $b = \pm\frac{\sqrt{3}}{2}$. Therefore, the vector we are looking for is $\frac{1}{2}\mathbf{i} \pm \frac{\sqrt{3}}{2}\mathbf{j}$.

(e) From the equation for the line $x + y = 0$, we see that the relation between the x and y component of the vector must be $x = -y$, so the vector looks like $\mathbf{v} = a\mathbf{i} - a\mathbf{j}$. To make this a unit vector, the length must be 1, so we have $\sqrt{a^2 + (-a)^2} = 1$, with solution $a = \pm\frac{\sqrt{2}}{2}$. Our vector must be $\pm(\frac{\sqrt{2}}{2}\mathbf{i} - \frac{\sqrt{2}}{2}\mathbf{j})$.

11. In terms of $\mathbf{i}$ and $\mathbf{j}$, determine the vector represented by the arrow extending from the origin to the midpoint of the line segment joining (1,4) with (3,8).
    Solution: The midpoint of the segment has components that are the averages of the individual components of the endpoints of the segment, so the midpoint is $(\frac{1+3}{2}, \frac{4+8}{2}) = (2, 6)$, and the vector from the origin to this point has components that are the differences of the points at the head and tail of the vector, or $(2 - 0)\mathbf{i} + (6 - 0)\mathbf{j} = 2\mathbf{i} + 6\mathbf{j}$.

13. If $\mathbf{V}$ is a unit vector in the xy plane making an angle of 30° with the positive y axis, express $\mathbf{V}$ in terms of $\mathbf{i}$ and $\mathbf{j}$.
    Solution: If $\mathbf{V}$ makes an angle of 30° with the positive y axis, it must make an angle of 120° with the positive x axis. Therefore the components of the vector will be the cosine and sine of 120°, and $\mathbf{V} = -\frac{1}{2}\mathbf{i} + \frac{\sqrt{3}}{2}\mathbf{j}$.

1.5 Space Vectors

In problems 1, 3, 5, 7, let $\mathbf{A} = 3\mathbf{i} + 4\mathbf{j}$, $\mathbf{B} = 2\mathbf{i} + 2\mathbf{j} - \mathbf{k}$ and $\mathbf{C} = 3\mathbf{i} - 4\mathbf{k}$.

1. Find $|\mathbf{A}|$, $|\mathbf{B}|$ and $|\mathbf{C}|$.
   Solution: As before, the magnitude (length) of a vector is the positive square root of the squares of the sums of its components. Thus, $|\mathbf{A}| = \sqrt{3^2 + 4^2} = \sqrt{25} = 5$, $|\mathbf{B}| = \sqrt{4 + 4 + 1} = 3$ and $|\mathbf{C}| = 5$

3. Determine $|\mathbf{A} - \mathbf{C}|$.
   Solution: $\mathbf{A} - \mathbf{C} = 4\mathbf{j} + 4\mathbf{k}$ and its magnitude is $\sqrt{16 + 16} = 4\sqrt{2}$.

5. Find the unit vector having the same direction as $\mathbf{A}$.
   Solution: We just divide $\mathbf{A}$ by its magnitude 5 to get $\mathbf{A}/|\mathbf{A}| = \frac{1}{5}(3\mathbf{i} + 4\mathbf{j})$.

7. Let α denote the angle between $\mathbf{A}$ and the positive x direction. Determine $\cos\alpha$.
   Solution: The cosine of the angle of the vector $\mathbf{A}$ with the x axis will be the x component of the unit vector $\mathbf{A}/|\mathbf{A}|$, or $\frac{3}{5}$.

9. Compute $|\mathbf{i} + \mathbf{j} + \mathbf{k}|$.
   Solution: $|\mathbf{i} + \mathbf{j} + \mathbf{k}| = \sqrt{3}$.

11. Write down the vector represented by the directed line segment OP, if O is the origin and $P(x, y, z)$ is a general point in space.
    Solution: $\mathbf{OP} = (x - 0)\mathbf{i} + (y - 0)\mathbf{j} + (z - 0)\mathbf{k} = x\mathbf{i} + y\mathbf{j} + z\mathbf{k}$.

1.6. TYPES OF VECTORS

13. What are the direction cosines of the vector $2\mathbf{i} - 2\mathbf{j} + \mathbf{k}$?
   Solution: The direction cosines are the x, y, and z components of the unit vector in the same direction, or $\frac{2}{3}$, $-\frac{2}{3}$, $\frac{1}{3}$.

15. Give a geometrical description of the locus of all points P for which OP represents a vector with direction cosine $\cos \alpha = 1/2$ (O is the origin).
   Solution: This is a cone symmetric about the x axis.

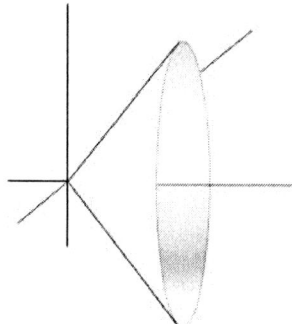

Figure 1.1: A cone symmetric about the x axis with angle $\pi/3$.

17. $\mathbf{A}$ is a vector with direction cosines $\cos \alpha$, $\cos \beta$ and $\cos \gamma$, respectively. What are the direction cosines of the reflected image of $\mathbf{A}$ in the yz plane? (Think of the yz plane as a mirror.)
   Solution: Reflection through the yz plane will not change the y and z components of a vector, but will reverse the sign of the x component, so $-\cos \alpha$, $\cos \beta$, and $\cos \gamma$.

19. Verify the commutative and associative laws of addition for space vectors by expressing them componentwise.
   Solution: Let $\mathbf{A} = a_1\mathbf{i} + a_2\mathbf{j} + a_3\mathbf{k}$, $\mathbf{B} = b_1\mathbf{i} + b_2\mathbf{j} + b_3\mathbf{k}$, and $\mathbf{C} = c_1\mathbf{i} + c_2\mathbf{j} + c_3\mathbf{k}$. Then

$$\begin{aligned}\mathbf{A} + \mathbf{B} &= (a_1\mathbf{i} + a_2\mathbf{j} + a_3\mathbf{k}) + (b_1\mathbf{i} + b_2\mathbf{j} + b_3\mathbf{k}) \\ &= (a_1 + b_1)\mathbf{i} + (a_2 + b_2)\mathbf{j} + (a_3 + b_3)\mathbf{k} \\ &= (b_1 + a_1)\mathbf{i} + (b_2 + a_2)\mathbf{j} + (b_3 + a_3)\mathbf{k} \\ &= (b_1\mathbf{i} + b_2\mathbf{j} + b_3\mathbf{k}) + (a_1\mathbf{i} + a_2\mathbf{j} + a_3\mathbf{k}) \\ &= \mathbf{B} + \mathbf{A},\end{aligned}$$

proving commutativity. To prove associativity,

$$\begin{aligned}(\mathbf{A} + \mathbf{B}) + \mathbf{C} &= ((a_1\mathbf{i} + a_2\mathbf{j} + a_3\mathbf{k}) + (b_1\mathbf{i} + b_2\mathbf{j} + b_3\mathbf{k})) + (c_1\mathbf{i} + c_2\mathbf{j} + c_3\mathbf{k}) \\ &= ((a_1 + b_1)\mathbf{i} + (a_2 + b_2)\mathbf{j} + (a_3 + b_3)\mathbf{k}) + (c_1\mathbf{i} + c_2\mathbf{j} + c_3\mathbf{k}) \\ &= (a_1 + b_1 + c_1)\mathbf{i} + (a_2 + b_2 + c_2)\mathbf{j} + (a_3 + b_3 + c_3)\mathbf{k} \\ &= (a_1\mathbf{i} + a_2\mathbf{j} + a_3\mathbf{k}) + (b_1 + c_1)\mathbf{i} + (b_2 + c_2)\mathbf{j} + (b_3 + c_3)\mathbf{k} \\ &= \mathbf{A} + (\mathbf{B} + \mathbf{C}).\end{aligned}$$

21. Why is it impossible for a vector to have direction angles $\alpha = 30°$ and $\beta = 30°$? Answer this both geometrically and in terms of the constraint *eq.* (1.5).
   Solution: To solve this geometrically, suppose $\gamma = 0$, placing the vector in the first quadrant of the xy plane. Then the direction cosines must add to $90°$, which they don't. In fact, we don't have to suppose $\gamma = 0$; we can just consider the projection of the vector into any coordinate plane, in this case the xy plane. To solve algebraically, note that $\cos 30 = \frac{\sqrt{3}}{2}$. Then $\cos^2 \alpha + \cos^2 \beta + \cos^2 \gamma = \frac{3}{4} + \frac{3}{4} + \cos^2 \gamma = 1$, which implies $\cos^2 \gamma = -\frac{\sqrt{1}}{2}$.

1.6 Types of Vectors

1. A particle moves from $(3, 7, 8)$ to $(5, 2, 0)$. Write its displacement in terms of $\mathbf{i}$, $\mathbf{j}$ and $\mathbf{k}$.
   Solution: We can think of points in space in two ways, as points (a, b, c) or as components of the

position vector $\mathbf{R} = a\mathbf{i} + b\mathbf{j} + c\mathbf{k}$. Therefore we can say the displacement vector is the difference of the final and initial points $\mathbf{v} = (5, 2, 0) - (3, 7, 8) = (5-3)\mathbf{i}(2-7)\mathbf{j}(0-8)\mathbf{k} = 2\mathbf{i} - 5\mathbf{j} - 8\mathbf{k}$.

3. The position vector of a moving particle at time t is $\mathbf{R} = 3\mathbf{i} + 4t^2\mathbf{j} - t^3\mathbf{k}$. Find its displacement during the time interval from $t = 1$ to $t = 3$.
   Solution: The displacement will be the difference of the position vectors $\mathbf{R}(3)$ and $\mathbf{R}(1)$. Therefore the displacement is $(3\mathbf{i} + 36\mathbf{j} - 27\mathbf{k}) - (3\mathbf{i} + 4\mathbf{j} - \mathbf{k}) = 32\mathbf{j} - 26\mathbf{k}$.

5. Strings are tied to a small metal ring and, by an arrangement of pulleys and weights, four forces are exerted on the ring. One force is directed upward with magnitude 3 lb, another is directed east with magnitude 6 lb, and a third is directed north with a magnitude 2 lb. The ring is in equilibrium (i.e., it is not moving). What is the magnitude of the fourth force that is counterbalancing the other three?
   Solution: A force can be written as a vector pointing in the direction of the force, with a length proportional to the magnitude of the force. Sums of vectors will then be sums of forces. Taking up as the positive z direction, north as positive y, south as negative y, east as positive x and west as negative x, write the vector sum of the forces as $\mathbf{F} = 3\mathbf{k} + 6\mathbf{i} + 2\mathbf{j}$. For the forces to be balanced, the vector sum of the forces must be zero, so we must add to this $\hat{\mathbf{F}} = -3\mathbf{k} - 6\mathbf{i} - 2\mathbf{j}$, which has magnitude $\sqrt{(-3)^2 + (-6)^2 + (-2)^2} = \sqrt{9 + 36 + 4} = \sqrt{49} = 7$ lbs.

7. Suppose a particle of electrical charge q_1 is located at $\mathbf{R}_1$, and q_2 is located at $\mathbf{R}_2$. the *Coulomb force* on particle 1 due to particle 2 is proportional to q_1 and q_2, and inversely proportional to the square of the distance between them; it is directed along the line from q_2 to q_1. Write down a vector formula for this force.
   Solution: The phrase "proportional to q_1 and q_2" means "equals a constant times the product of q_1 and q_2", or $k_1 q_1 q_2$. This is common wording, but the meaning is not very clear. The phrase "inversely proportional to the square of the distance between them" is more descriptive, and means $\frac{k_2}{|\mathbf{R}_1 - \mathbf{R}_2|^2}$. The phrase "directed along the line from q_2 to q_1" means "in the direction of the *unit vector* $\frac{\mathbf{R}_1 - \mathbf{R}_2}{|\mathbf{R}_1 - \mathbf{R}_2|}$", since q_1 is at $\mathbf{R}_1$ and q_2 is at $\mathbf{R}_2$. Altogether, the force is the products of these proportions and the direction vector, or $\mathbf{C} = \frac{kq_1q_2}{|\mathbf{R}_1-\mathbf{R}_2|^2} \frac{\mathbf{R}_1-\mathbf{R}_2}{|\mathbf{R}_1-\mathbf{R}_2|} = \frac{kq_1q_2}{|\mathbf{R}_1-\mathbf{R}_2|^3}(\mathbf{R}_1 - \mathbf{R}_2)$, where we have combined the proportionality constants into one.

1.7 Some Problems in Geometry

1. Find the angle between $2\mathbf{i} + \mathbf{j} + 2\mathbf{k}$ and $3\mathbf{i} - 4\mathbf{k}$.
   Solution: Because $\mathbf{A} \cdot \mathbf{B} = |\mathbf{A}||\mathbf{B}|\cos\theta$, $\theta = \cos^{-1}\frac{\mathbf{A}\cdot\mathbf{B}}{|\mathbf{A}||\mathbf{B}|}$, so the angle between $2\mathbf{i}+\mathbf{j}+2\mathbf{k}$ and $3\mathbf{i}-4\mathbf{k}$ is $\cos^{-1}\frac{-2}{15}$.

3. Find the three angles of the triangle with vertices (2,-1,1),(1,-3,-5),(3,-4,-4).
   Solution: Use the same method as in the previous problem, making sure you subtract the points in the proper order to get the appropriate vectors. In each case, the appropriate pair of vectors is obtained by subtracting each point in turn from the remaining pairs, so we find the angle between the pairs:

   $$(2,-1,1) - (1,-3,-5) = \mathbf{i} + 2\mathbf{j} + 6\mathbf{k}$$
   and $$(3,-4,-4) - (1,-3,-5) = 2\mathbf{i} - \mathbf{j} + \mathbf{k},$$
   which is $\cos^{-1}\frac{(\mathbf{i}+2\mathbf{j}+6\mathbf{k})\cdot(2\mathbf{i}-\mathbf{j}+\mathbf{k})}{|\mathbf{i}+2\mathbf{j}+6\mathbf{k}||2\mathbf{i}-\mathbf{j}+\mathbf{k}|} = \cos^{-1}\frac{6}{\sqrt{41}\sqrt{6}}$

   $$(1,-3,-5) - (2,-1,1) = -\mathbf{i} - 2\mathbf{j} - 6\mathbf{k}$$
   and $$(3,-4,-4) - (2,-1,1) = \mathbf{i} - 3\mathbf{j} - 5\mathbf{k},$$
   which is $\cos^{-1}\frac{(-\mathbf{i}-2\mathbf{j}-6\mathbf{k})\cdot(\mathbf{i}-3\mathbf{j}-5\mathbf{k})}{|-\mathbf{i}-2\mathbf{j}-6\mathbf{k}||\mathbf{i}-3\mathbf{j}-5\mathbf{k}|} = \cos^{-1}\frac{35}{\sqrt{41}\sqrt{35}}$

1.7. SOME PROBLEMS IN GEOMETRY

$$(2, -1, 1) - (3, -4, -4) = -\mathbf{i} + 3\mathbf{j} + 5\mathbf{k}$$

and $\quad (1, -3, -5) - (3, -4, -4) = -2\mathbf{i} + \mathbf{j} - \mathbf{k},$

which is $\quad \cos^{-1} \frac{(-\mathbf{i}+3\mathbf{j}+5\mathbf{k})\cdot(-2\mathbf{i}+\mathbf{j}-\mathbf{k})}{|-\mathbf{i}+3\mathbf{j}+5\mathbf{k}||-2\mathbf{i}+\mathbf{j}-\mathbf{k}|} = \cos^{-1} \frac{0}{\sqrt{35}\sqrt{6}}.$

5. Show that $\mathbf{i}+\mathbf{j}+\mathbf{k}$ is perpendicular to the plane $x+y+z = 0$. (*Hint*: This plane passes through the origin. Show that $\mathbf{i}+\mathbf{j}+\mathbf{k}$ is perpendicular to every vector extending from the origin to a point in the plane.)
Solution: The position vector $x\mathbf{i} + y\mathbf{j} + z\mathbf{k}$ has its tip in the plane if $x+y+z = 0$. The scalar (dot) product of two vectors is zero if they are perpendicular, and the scalar product of $\mathbf{i}+\mathbf{j}+\mathbf{k}$ and is $x+y+z$, which is zero precisely when $x\mathbf{i}+y\mathbf{j}+z\mathbf{k}$ is in the plane, so the plane $x+y+z = 0$ is perpendicular to $\mathbf{i}+\mathbf{j}+\mathbf{k}$.

7. Using vector methods, prove directly that if two sides of a quadrilateral are parallel and equal in magnitude, the other two sides are also.
Solution: Follow the perimeter of the quadrilateral and label the successive vertices (a_1, b_1, c_1), (a_2, b_2, c_2), (a_3, b_3, c_3), (a_4, b_4, c_4). Then two opposite sides are given by the free vectors $(a_1 - a_2)\mathbf{i} + (b_1 - b_2)\mathbf{j} + (c_1 - c_2)\mathbf{k}$ and $(a_4 - a_3)\mathbf{i} + (b_4 - b_3)\mathbf{j} + (c_4 - c_3)\mathbf{k}$. If these sides are parallel and equal in length, then their components must be equal, so $a_1 - a_2 = a_4 - a_3$, $b_1 - b_2 = b_4 - b_3$, and $c_1 - c_2 = c_4 - c_3$ or $a_1 - a_4 = a_2 - a_3$, $b_1 - b_4 = b_2 - b_3$, and $c_1 - c_4 = c_2 - c_3$. But this is just the condition that shows the other pair of sides $(a_1 - a_4)\mathbf{i} + (b_1 - b_4)\mathbf{j} + (c_1 - c_4)\mathbf{k}$ and $(a_2 - a_3)\mathbf{i} + (b_2 - b_3)\mathbf{j} + (c_2 - c_3)\mathbf{k}$ are parallel, so we have proved it.

9. Show that the diagonals of a parallelogram bisect each other.
Solution: There are several ways to do this, including the suggestion in the back of the book to copy the method of example 1.3. Instead, let's use the vectors and relations among their components we found in problem 7. The two diagonals are between the points (a_1, b_1, c_1) and (a_3, b_3, c_3) and the points (a_2, b_2, c_2) and (a_4, b_4, c_4). We can interpolate between (a_1, b_1, c_1) and (a_3, b_3, c_3) by $(1-t)(a_1, b_1, c_1) + t(a_3, b_3, c_3)$ and between (a_2, b_2, c_2) and (a_4, b_4, c_4) by $(1-t)(a_2, b_2, c_2) + t(a_4, b_4, c_4)$ for $0 \le t \le 1$. Halfway between (a_1, b_1, c_1) and (a_3, b_3, c_3) is $\frac{1}{2}(a_1 + a_3, b_1 + b_3, c_1 + c_3)$ and halfway between (a_2, b_2, c_2) and (a_4, b_4, c_4) is $\frac{1}{2}(a_2 + a_4, b_2 + b_4, c_2 + c_4)$. But by the relation found in problem 7, these are identical points, so the diagonals bisect each other.

11. A treasure map has n villages marked on it, and it contains the following instructions. Start at village A, go 1/2 of the way to village B, 1/3 of the way to village C, 1/4 of the way to village D, and so forth. The treasure is buried at the last stop. *Problem*: You lose the instructions, and don't know in what order to select the villages. Show that it doesn't matter! (see Figure 1.2) Then relate this to example 1.4 for $n = 3$.
Solution: Let $\mathbf{R}_1, \mathbf{R}_2, ..., \mathbf{R}_n$ be the position vectors locating the villages. Then our position at any time is given by a vector sum, and our positions just before changing directions are
$\mathbf{R}_1$
$\mathbf{R}_1 + \frac{1}{2}(\mathbf{R}_2 - \mathbf{R}_1)$
$\mathbf{R}_1 + \frac{1}{2}(\mathbf{R}_2 - \mathbf{R}_1) + \frac{1}{3}(\mathbf{R}_3 - (\mathbf{R}_1 + \frac{1}{2}(\mathbf{R}_2 - \mathbf{R}_1)))$
$\mathbf{R}_1 + \frac{1}{2}(\mathbf{R}_2 - \mathbf{R}_1) + \frac{1}{3}(\mathbf{R}_3 - (\mathbf{R}_1 + \frac{1}{2}(\mathbf{R}_2 - \mathbf{R}_1)))$
$+ \frac{1}{4}(\mathbf{R}_4 - (\mathbf{R}_1 + \frac{1}{2}(\mathbf{R}_2 - \mathbf{R}_1) + \frac{1}{3}(\mathbf{R}_3 - (\mathbf{R}_1 + \frac{1}{2}(\mathbf{R}_2 - \mathbf{R}_1)))))$
and so on. Now calculate each of these sums. We get $\mathbf{R}_1$, $\frac{1}{2}(\mathbf{R}_1 + \mathbf{R}_2)$, $\frac{1}{3}(\mathbf{R}_1 + \mathbf{R}_2 + \mathbf{R}_3)$ and $\frac{1}{4}(\mathbf{R}_1 + \mathbf{R}_2 + \mathbf{R}_3 + \mathbf{R}_4)$. Notice that in each case, the order of the subscripts does not matter – the villages can be visited in any order!

To prove this in all generality we prove by induction. We use proof by induction when we have a sequence of equations that depend explicitly on their order the sequence, as in this case. There are two steps: first, we propose that the solution has a particular form depending on an index assigning its place in the sequence, and give a specific example, usually one of the first formulas in the sequence. Second, we show that if the solution has that form for a particular index, then it holds for the next higher index. Because we have shown the formula true for some small index, and that if it is true for some index then it is true for the next index, we know it is true for all indices.

For this problem we propose that our position at step n is $\frac{1}{n}\sum_{i=1}^{n} \mathbf{R}_i$, and that the next step $n+1$ our position is $\frac{1}{n}\sum_{i=1}^{n} \mathbf{R}_i + \frac{1}{n+1}(\mathbf{R}_{n+1} - \frac{1}{n}\sum_{i=1}^{n} \mathbf{R}_i)$. This is certainly true for each step we have calculated so far, so it holds for one value of n in particular. Now we show that if it holds for n, it holds for $n+1$. This means that at $n+1$ the pattern must be $\frac{1}{n+1}\sum_{i=1}^{n+1} \mathbf{R}_i$. We know that our position at time $n+1$ is $\frac{1}{n}\sum_{i=1}^{n} \mathbf{R}_i + \frac{1}{n+1}(\mathbf{R}_{n+1} - \frac{1}{n}\sum_{i=1}^{n} \mathbf{R}_i)$. Multiplying

through by $\frac{1}{n+1}$ and combining the two summations, we have $\left(\frac{1}{n} - \frac{1}{n(n+1)}\right) \sum_{i=0}^{n} \mathbf{R}_i + \frac{1}{n+1}\mathbf{R}_{n+1}$
$= \frac{1}{n+1}\sum_{i=0}^{n} \mathbf{R}_i + \frac{1}{n+1}\mathbf{R}_{n+1} = \frac{1}{n+1}\sum_{i=0}^{n+1} \mathbf{R}_i$, and we're done.

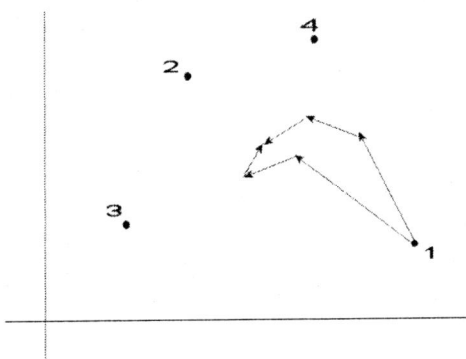

Figure 1.2: Beginning at village 1, we go halfway to 2, one third of the way to 3, one fourth of the way to 4, and so on. If we go halfway to village 4, then a third of the way to 2 and one forth of the way to 3, the result is the same.

13. True or false: $3x - 4y + 5z = 0$ represents a plane passing through the origin.
    Solution: True, because $x = 0$, $y = 0$, $z = 0$ satisfies the equation of the plane so the origin is included in the plane.

15. True or false: The locus of points for which $x = 3$ and $y = 4$ is a line parallel to the z axis whose distance from the z axis is 5.
    Solution: True, because $d = \sqrt{(x-0)^2 + (y-0)^2} = \sqrt{3^2 + 4^2} = 5$.

17. Write down the equation of a sphere centered at the point $(2, 3, 4)$ having radius 3.
    Solution: $(x-2)^2 + (y-3)^2 + (z-4) = 3^2$.

19. Do the equations $x = y = z$ represent a line or a plane?
    Solution: They represent a line because it is in the form of an equation of a line through the origin $\frac{x-0}{1} = \frac{y-0}{1} = \frac{z-0}{1}$.

21. What is the locus of points for which $(x-2)^2 + (y+3)^2 + (z-4)^2 = 0$?
    Solution: This is the point $(2, -3, 4)$, because it is a sphere centered at that point with radius zero.

23. What is the distance between the points $(2, 3, 4)$ and $(5, 3, 8)$?
    Solution: The distance is the 5, the length of the vector $3\mathbf{i} + 4\mathbf{k}$ between the points.

25. What is the distance between the point $(0, 3, 0)$ and the cylinder $x^2 + y^2 = 4$ (You probably won't find a formula for this in any of your books. Just use some common sense.)
    Solution: The cylinder is a distance 2 from the z axis, and the point $(0, 3, 0)$ is a distance 3 away, so the distance between the point and the cylinder is 1.

27. Do you know what figure is represented by the equation $(x/2)^2 + (y/3)^2 + (z/4)^2 = 1$? (If so, you know more analytic geometry than is required to read this book.)
    Solution: This is an ellipsoid.

1.8 Equations of a Line

1. Find parametric equations of the line passing through the origin parallel to $3\mathbf{i} - 2\mathbf{j} + 7\mathbf{k}$.
    Solution: $(0\mathbf{i} + 0\mathbf{j} + 0\mathbf{k}) + t(3\mathbf{i} - 2\mathbf{j} + 7\mathbf{k}) = 3t\mathbf{i} - 2t\mathbf{j} + 7t\mathbf{k}$ or $x = 3t$, $y = -2t$, $z = 7t$.

1.8. EQUATIONS OF A LINE

3. Find the equations of the line parallel to the z axis passing through the point $(1, 2, 3)$.
Solution: There are two types of equations for a line – parametric equations and non-parametric equations. The parametric equations are $x = 1$, $y = 2$, $z = 3 + t$. The non-parametric equations are obtained by eliminating t from the parametric equations. However, in this case there is no t to eliminate from the first two. If we rewrite the parametric equations as $x = 1 + 0t$, $y = 2 + 0t$, $z = 3 + t$ and eliminate t, we obtain $\frac{x-1}{0} = \frac{y-2}{0} = \frac{z-3}{1}$ which is nonsense, so we stick with the parametric notation.

5. Find two unit vectors parallel to the line $x = 2y = 3z + 3$. These equations can be written in form (1.10) as follows: $x = \frac{y}{\frac{1}{2}} = \frac{z+1}{\frac{1}{3}}$
Solution: There are always two distinct unit vectors parallel to any line, and one is the negative of the other. Writing $x = \frac{y}{\frac{1}{2}} = \frac{z+1}{\frac{1}{3}}$ as $\frac{x-0}{1} = \frac{y-0}{\frac{1}{2}} = \frac{z-(-1)}{\frac{1}{3}}$ it is easier to see that $(x_0, y_0, z_0) = (0, 0, -1)$ and $(a, b, c) = (1, 1/2, 1/3)$. To make a unit vector in the direction of $a\mathbf{i} + b\mathbf{j} + c\mathbf{k}$, it is a little easier to multiply $\mathbf{i} + \frac{1}{2}\mathbf{j} + \frac{1}{3}\mathbf{k}$ by 6 then normalize: $\mathbf{v} = 6\mathbf{i} + 3\mathbf{j} + 2\mathbf{k}$ with length $\sqrt{6^2 + 3^2 + 2^2} = 7$, so the unit vector we seek is $\frac{6}{7}\mathbf{i} + \frac{3}{7}\mathbf{j} + \frac{2}{7}\mathbf{k}$.

7. Find equations of the line passing through the origin and parallel to the line $x - 3 = \frac{y+2}{4} = 1 - z$
Solution: The equation of a line passing through a point and parallel to a given line is the sum of the position vector to that point plus a parameter, say t, times a vector $\mathbf{v}$ in the direction of the line. In this case the position vector to the origin is $\mathbf{R} = 0\mathbf{i} + 0\mathbf{j} + 0\mathbf{k}$ and a direction vector is found by taking a specific value (any value will do) of t in the parametric equation $x = t + 3$, $y = 4t - 2$, $z = -t + 1$. Let $t = 0$, for instance. Then $\mathbf{v} = 3\mathbf{i} - 2\mathbf{j} + \mathbf{k}$, and the line is given by $L(t) = t(3\mathbf{i} - 2\mathbf{j} + \mathbf{k})$. There are always an infinite number of formulas for any line, because the direction vector can be any length. The only practical difference between these formulas is which point of the line we are on for which value of t.

9. Find equations of the line passing through the points $(1, 4, -1)$ and $(2, 2, 7)$.
Solution: The direction vector $\mathbf{v} = \mathbf{i} - 2\mathbf{j} + 8\mathbf{k}$ is found by taking the difference of the two points, and either point can serve as a "base point". One equation for the line is $(\mathbf{i} + 4\mathbf{j} - \mathbf{k}) + t(\mathbf{i} - 2\mathbf{j} + 8\mathbf{k})$. We can write this in parametric form as $x = 1 + t$, $y = 4 - 2t$, $z = -1 + 8t$. The non-parametric form is $\frac{x-1}{1} = \frac{y-4}{-2} = \frac{z+1}{8}$.

11. Find the angle between the two intersecting lines

$$\frac{x-1}{3} = \frac{y-3}{4} = \frac{z}{5} \text{ and } \frac{x-1}{2} = 3 - y = 2z$$

Solution: To find the angle between lines, we need to find vectors $\mathbf{u}$ and $\mathbf{v}$ in the directions of those lines. Then we can use the formula $\mathbf{u} \cdot \mathbf{v} = |\mathbf{u}||\mathbf{v}|\cos\theta$ to find the angle θ between them. We can read off vectors directly from the non-parametric equations, but first let's systematically determine the vectors by rewriting the line equations as parametric equations and see what the trick is. The parametric equations are $x = 3t + 1$, $y = 4t + 3$, $z = 5t$ and $x = 2t + 1$, $y = 3 - t$, $z = \frac{1}{2}t$. Then vectors in the directions of these lines can be found by taking the difference between the equations for two different values of t, say $t = 0$ and $t = 1$, because this will be the difference of two different points on the line. Notice that for $t = 0$, the lines intersect. The difference for the first line is $x(1) - x(0) = 3$, $y(1) - y(0) = 4$, $z(1) - z(0) = 5$, and for the second line $x(1) - x(0) = 2$, $y(1) - y(0) = -1$, $z(1) - z(0) = \frac{1}{2}$. Notice that the coefficient of t are all that remain after these calculations, and that these coefficients are the denominators of the non-parametric equations (when they are rewritten as $\frac{x-1}{3} = \frac{y-3}{4} = \frac{z}{5}$ and $\frac{x-1}{2} = \frac{y-3}{-1} = \frac{z-0}{1/2}$ to have denominators). Writing these as vectors $\mathbf{u} = 3\mathbf{i} + 4\mathbf{j} + 5\mathbf{k}$ and $\mathbf{v} = 2\mathbf{i} - \mathbf{j} + \frac{1}{2}\mathbf{k}$, we calculate $\cos^{-1}\frac{\mathbf{u} \cdot \mathbf{v}}{|\mathbf{u}||\mathbf{v}|} = \cos^{-1}\frac{9/2}{5/2\sqrt{42}} = \cos^{-1}\frac{3\sqrt{42}}{70}$

13. Solve exercise 9 by making use of *eq.(1.11)*.
Solution: Equation 1.11 is a linear interpolation between two points $(1-t)p_1 + tp_2$. The points in exercise 9 are $(1, 4, -1)$ and $(2, 2, 7)$, so our line is $L(t) = (1-t)(\mathbf{i} + 4\mathbf{j} - \mathbf{k}) + t(2\mathbf{i} + 2\mathbf{j} + 7\mathbf{k}) = (1+t)\mathbf{i} + (4-2t)\mathbf{j} + (8t-1)\mathbf{k}$.

15. Find the point(s) of intersection of the following pairs of straight lines:

(a) $\mathbf{R} = (5\mathbf{i} + 4\mathbf{j} + 5\mathbf{k})t + 7\mathbf{i} + 6\mathbf{j} + 8\mathbf{k}$ and $\mathbf{R} = (6\mathbf{i} + 4\mathbf{j} + 6\mathbf{k})t + 8\mathbf{i} + 6\mathbf{j} + 9\mathbf{k}$
(b) $\mathbf{R} = (3\mathbf{i} + 2\mathbf{j} + \mathbf{k})t + 2\mathbf{k}$ and $\mathbf{R} = (6\mathbf{i} + 4\mathbf{j} + 2\mathbf{k})t + 3\mathbf{i} + 2\mathbf{j} + 1\mathbf{k}$

(c) $\mathbf{R} = (3\,\mathbf{i} - \mathbf{j} + \mathbf{k})t$ and $\mathbf{R} = (-6\,\mathbf{i} + 2\,\mathbf{j} - 2\,\mathbf{k})t + 2\,\mathbf{i}$

(d) $\mathbf{R} = (\mathbf{i} + \mathbf{j} + \mathbf{k})t$ and $\mathbf{R} = (\mathbf{i} + \mathbf{j} - 3\,\mathbf{k})t - \mathbf{i} + \mathbf{j}$

Solution: To find the intersection of two lines given in parametric form, we must solve three pairs of simultaneous equations in two parameters. We have no guarantee that the ts will be the same at the points of intersections of these lines.

(a) The formulas for the lines can be rewritten as
$\mathbf{R} = (7 + 5t)\,\mathbf{i} + (6 + 4t)\,\mathbf{j} + (8 + 5t)\,\mathbf{k}$ and
$\mathbf{R} = (8 + 6s)\,\mathbf{i} + (6 + 4s)\,\mathbf{j} + (9 + 6s)\,\mathbf{k}$, so we see we need to solve $(7 + 5t) = (8 + 6s)$, $(6 + 4t) = (6 + 4s)$ and $(8 + 5t) = (9 + 6s)$ simultaneously. We see from the second equation that $t = s$, and either the first or third gives us that $t = s = -1$, so the point of intersection is $(2, 2, 3)$.

(b) Writing $\mathbf{R} = 3t\,\mathbf{i} + 2t\,\mathbf{j} + (2 + t)\,\mathbf{k}$ and $\mathbf{R} = (3 + 6s)\,\mathbf{i} + (2 + 4s)\,\mathbf{j} + (3 + 2s)\,\mathbf{k}$, we see that we must solve $3t = 3 + 6s$, $2t = 2 + 4s$ and $2 + t = 3 + 2s$. All three equations show that $t = 1 + 2s$, and substituting this into the first equation gives us the second, so the lines are identical: they coincide.

(c) For $\mathbf{R} = 3t\,\mathbf{i} - t\,\mathbf{j} + t\,\mathbf{k}$ and $\mathbf{R} = (2 - 6s)\,\mathbf{i} + 2s\,\mathbf{j} - 2s\,\mathbf{k}$ to intersect, we must have $3t = 2 - 6s$, $-t = 2s$, and $t = -2s$. These have no simultaneous solution, so the lines do not intersect.

(d) If $\mathbf{R} = t\,\mathbf{i} + t\,\mathbf{j} + t\,\mathbf{k}$ and $\mathbf{R} = (s - 1)\,\mathbf{i} + (s + 1)\,\mathbf{j} - 3s\,\mathbf{k}$ intersect, then we must have $t = s - 1$ and $t = s + 1$, so there is no point of intersection.

17. Equations (1.9) seem to indicate that six numbers (x_0, y_0, z_0, a, b, c) are needed to specify a straight line. Is this correct?
Solution: In Euclidean three-space this is certainly true in general.

1.9 Scalar Products

1. Find the scalar product of $3\,\mathbf{i} + 8\,\mathbf{j} - 2\,\mathbf{k}$ with $5\,\mathbf{i} + \mathbf{j} + 2\,\mathbf{k}$.
Solution: $(3\,\mathbf{i} + 8\,\mathbf{j} - 2\,\mathbf{k}) \cdot (5\,\mathbf{i} + \mathbf{j} + 2\,\mathbf{k}) = 3 \times 5 + 8 \times 1 - 2 \times 2 = 19$

3. Find the scalar product of $3\,\mathbf{i} + 4\,\mathbf{j}$ with $5\,\mathbf{j} - 10\,\mathbf{k}$.
Solution: $(3\,\mathbf{i} + 4\,\mathbf{j}) \cdot (5\,\mathbf{j} - 10\,\mathbf{k}) = 3 \times 0 + 4 \times 5 + 0 \times 10 = 20$

5. Find the angle between $2\,\mathbf{i}$ and $3\,\mathbf{i} + 4\,\mathbf{j}$.
Solution: Using $\mathbf{u} \cdot \mathbf{v} = |\mathbf{u}||\mathbf{v}|\cos\theta$, we find $\theta = \cos^{-1} \frac{\mathbf{u}\cdot\mathbf{v}}{|\mathbf{u}||\mathbf{v}|} = \cos^{-1} \frac{2\times 3 + 4\times 0}{2\times 5} = \cos^{-1} \frac{3}{5}$.

7. Find the component of $8\,\mathbf{i} + \mathbf{j}$ in the direction of $\mathbf{i} + 2\,\mathbf{j} - 2\,\mathbf{k}$.
Solution: Recall that the parallel and perpendicular components of a vector $\mathbf{B}$ relative to $\mathbf{A}$ are $\mathbf{B}_{\parallel} = \frac{\mathbf{B}\cdot\mathbf{A}}{\mathbf{A}\cdot\mathbf{A}}\mathbf{A}$ and $\mathbf{B}_{\perp} = \mathbf{B} - \frac{\mathbf{B}\cdot\mathbf{A}}{\mathbf{A}\cdot\mathbf{A}}\mathbf{A}$. With $\mathbf{B} = 8\,\mathbf{i} + \mathbf{j}$ and $\mathbf{A} = \mathbf{i} + 2\,\mathbf{j} - 2\,\mathbf{k}$, we get $\mathbf{B}_{\parallel} = \frac{(8\,\mathbf{i}+\mathbf{j})\cdot(\mathbf{i}+2\,\mathbf{j}-2\,\mathbf{k})}{(\mathbf{i}+2\,\mathbf{j}-2\,\mathbf{k})\cdot(\mathbf{i}+2\,\mathbf{j}-2\,\mathbf{k})}(\mathbf{i} + 2\,\mathbf{j} - 2\,\mathbf{k}) = \frac{10}{9}(\mathbf{i} + 2\,\mathbf{j} - 2\,\mathbf{k})$, so the component in the $\mathbf{A}$ direction is $\frac{10}{9}|\mathbf{A}| = \frac{10}{3}$.

9. Find the component of the force $5\,\mathbf{i} + 7\,\mathbf{j} - \mathbf{k}$ in the direction of the displacement PQ, where $P(3, 0, 1)$ and $Q(4, 4, 4)$ are points in space.
Solution: Using the formula $|\mathbf{B}_{\parallel}| = \frac{\mathbf{B}\cdot\mathbf{A}}{\mathbf{A}\cdot\mathbf{A}}|\mathbf{A}|$ along with the displacement vector $\mathbf{A} = \mathbf{i} + 4\,\mathbf{j} + 3\,\mathbf{k}$ and $\mathbf{B} = 5\,\mathbf{i} + 7\,\mathbf{j} - \mathbf{k}$, we calculate $|\mathbf{B}_{\parallel}| = \frac{(5\,\mathbf{i}+7\,\mathbf{j}-\mathbf{k})\cdot(\mathbf{i}+4\,\mathbf{j}+3\,\mathbf{k})}{(\mathbf{i}+4\,\mathbf{j}+3\,\mathbf{k})\cdot(\mathbf{i}+4\,\mathbf{j}+3\,\mathbf{k})}|\mathbf{i} + 4\,\mathbf{j} + 3\,\mathbf{k}| = \frac{15\sqrt{26}}{13}$.

11. If $\mathbf{A} \cdot \mathbf{A} = 0$ and $\mathbf{A} \cdot \mathbf{B} = 0$, what can you conclude about the vector $\mathbf{B}$?
Solution: You can conclude nothing, because the first condition indicates that $\mathbf{A}$ is the zero vector.

13. Determine s and t so that $\mathbf{C} - s\mathbf{A} - t\mathbf{B}$ is perpendicular to both $\mathbf{A}$ and $\mathbf{B}$, given that $\mathbf{A} = \mathbf{i} + \mathbf{j} + 2\,\mathbf{k}$, $\mathbf{B} = 2\,\mathbf{i} - \mathbf{j} + \mathbf{k}$, $\mathbf{C} = 2\,\mathbf{i} - \mathbf{j} + 4\,\mathbf{k}$
Solution: We must solve the pair of equations $\mathbf{A} \cdot (\mathbf{C} - s\mathbf{A} - t\mathbf{B}) = 0$ and $\mathbf{B} \cdot (\mathbf{C} - s\mathbf{A} - t\mathbf{B}) = 0$. Combining these, we get $\mathbf{C} \cdot (\mathbf{A} - \mathbf{B}) - s\mathbf{A}(\mathbf{A} - \mathbf{B}) - t\mathbf{B}(\mathbf{A} - \mathbf{B}) = 0$. This is a polynomial in s and t, so the coefficients on each side of the equal sign must be equal. We see the constant coefficient is zero, and $s\mathbf{A}(\mathbf{A} - \mathbf{B}) + t\mathbf{B}(\mathbf{A} - \mathbf{B}) = 0$ indicates that $3s - 3t = 0$, so as long as $s = t$, the conditions will be satisfied.

1.9. SCALAR PRODUCTS

15. By interpreting $2x + 3y + 4z$ as a scalar product, show that $2\mathbf{i} + 3\mathbf{j} + 4\mathbf{k}$ is perpendicular to the plane $2x + 3y + 4z = 0$.
Solution: Let $\mathbf{R} = x\mathbf{i} + y\mathbf{j} + z\mathbf{k}$ be the position vector. Then when x, y, z are restricted to lie in a plane, any vector $\mathbf{v}$ for which $\mathbf{v} \cdot \mathbf{R} = 0$ will be perpendicular to that plane. Thus, $(2\mathbf{i} + 3\mathbf{j} + 4\mathbf{k}) \cdot (x\mathbf{i} + y\mathbf{j} + z\mathbf{k}) = 0$ is a plane for which $2\mathbf{i} + 3\mathbf{j} + 4\mathbf{k}$ is a normal, or perpendicular vector.

17. If $\mathbf{u}$ and $\mathbf{v}$ are unit vectors, and θ is the angle between them, find $1/2|\mathbf{u}-\mathbf{v}|$ in terms of θ.
Solution: Because $\mathbf{u}$ and $\mathbf{v}$ are unit vectors, $\frac{1}{2}|\mathbf{u}-\mathbf{v}|$ is half the base length of an isosceles triangle (check this by computing the angles between $\mathbf{u}$ and $\mathbf{v}$ and $\mathbf{u}-\mathbf{v}$) whose equal sides have length 1. Bisecting the vertex angle between the equal sides, we get a right triangle with hypotenuse length 1 and angle $\theta/2$. Thus one half the base length is $\sin\theta/2$

19. Prove, by vector methods, that the median from the vertex angle of an isosceles triangle is perpendicular to the base.
Solution: From the previous problem, we know that a vector in the direction of the base in an isosceles triangle with equal sides $\mathbf{u}$ and $\mathbf{v}$ is given by $\mathbf{u}-\mathbf{v}$. A vector that bisects the vertex angle is given by $\mathbf{u}+\mathbf{v}$ and the scalar product of these is $(\mathbf{u}-\mathbf{v}) \cdot (\mathbf{u}+\mathbf{v}) = \mathbf{u}\cdot\mathbf{u} - \mathbf{v}\cdot\mathbf{v} = 1 - 1 = 0$, so the bisector of the vertex angle is perpendicular to the base.

21. Prove the triangle inequality of section 1.3, $|\mathbf{A}+\mathbf{B}| \leq |\mathbf{A}|+|\mathbf{B}|$, for any vectors $\mathbf{A}$ and $\mathbf{B}$. *Hint:* Square both sides and use the scalar product.
Solution: Following the suggestion, we find $|\mathbf{A}+\mathbf{B}|^2 = (\mathbf{A}+\mathbf{B})\cdot(\mathbf{A}+\mathbf{B}) = \mathbf{A}\cdot\mathbf{A}+2\mathbf{B}\cdot\mathbf{A}+\mathbf{B}\cdot\mathbf{B} = \cos\theta(|\mathbf{A}||\mathbf{A}|+2|\mathbf{B}||\mathbf{A}|+|\mathbf{B}||\mathbf{B}|)$ and $(|\mathbf{A}|+|\mathbf{B}|)^2 = |\mathbf{A}||\mathbf{A}|+2|\mathbf{B}||\mathbf{A}|+|\mathbf{B}||\mathbf{B}|$. Because $\cos\theta \leq 1$, we have $|\mathbf{A}+\mathbf{B}| \leq |\mathbf{A}|+|\mathbf{B}|$.

23. Consider the cube in figure 1.27. Find the angles between

(a) the face diagonals AB and AC.

(b) the principal diagonal AD and the face diagonal AB.

(c) the principal diagonal AD and the edge AE.

Solution: Without loss of generality, we can place the cube so that A is at the origin and AE points along the positive x axis. Then $\mathbf{u} = \mathbf{i}$ is the vector from A to E, $\mathbf{v} = \mathbf{j}$ is the vector from E to B, and $\mathbf{w} = -\mathbf{k}$ is the vector from E to C. The vector from A to B is $\mathbf{u}+\mathbf{v} = \mathbf{i}+\mathbf{j}$ and has length $\sqrt{2}$, the vector from A to C is $\mathbf{u}+\mathbf{w} = \mathbf{i}-\mathbf{k}$ and has length $\sqrt{2}$, and the vector from A to D is $\mathbf{u}+\mathbf{w}+\mathbf{v} = \mathbf{i}+\mathbf{j}-\mathbf{k}$ and has length $\sqrt{3}$.

(a) The angle between the face diagonals AB and AC is the angle between $\mathbf{u}+\mathbf{v}$ and $\mathbf{u}+\mathbf{w}$ which is $\cos^{-1}\frac{(\mathbf{i}+\mathbf{j})\cdot(\mathbf{i}-\mathbf{k})}{\sqrt{2}\sqrt{2}} = \cos^{-1}\frac{1}{2} = \frac{\pi}{3}$.

(b) The angle between the principal diagonal AD and the face diagonal AB is the angle between $\mathbf{i}+\mathbf{j}-\mathbf{k}$ and $\mathbf{i}+\mathbf{j}$ which is $\cos^{-1}\frac{(\mathbf{i}+\mathbf{j}-\mathbf{k})\cdot(\mathbf{i}+\mathbf{j})}{\sqrt{2}\sqrt{3}} = \cos^{-1}\frac{2}{\sqrt{6}} = \cos^{-1}\frac{\sqrt{6}}{3}$.

(c) The angle between the principal diagonal AD and the edge AE is the angle between $\mathbf{i}+\mathbf{j}-\mathbf{k}$ and $\mathbf{i}$ which is $\cos^{-1}\frac{(\mathbf{i}+\mathbf{j}-\mathbf{k})\cdot(\mathbf{i})}{\sqrt{1}\sqrt{3}} = \cos^{-1}\frac{1}{\sqrt{3}} = \cos^{-1}\frac{\sqrt{3}}{3}$.

25. Prove: the diagonals of a rectangle are perpendicular if and only if the rectangle is a square.
Solution: Let $\mathbf{A}$ and $\mathbf{B}$ be the sides of a rectangle. Then the diagonals are given by $\mathbf{A}+\mathbf{B}$ and $\mathbf{A}-\mathbf{B}$. The scalar product of these diagonals is $\mathbf{A}\cdot\mathbf{A}-\mathbf{B}\cdot\mathbf{B}$, which is zero (i.e., they are perpendicular) only if $|\mathbf{A}|=|\mathbf{B}|$.

27. Prove: the sum of the squares of the sides of any quadrilateral, minus the sum of the squares of the two diagonals, equals four times the square of the distance between the midpoints of the diagonals.
Solution: Let p_1, p_2, p_3, p_4 be the vertices of the quadrilateral taken in order as we traverse its perimeter. Let the edges be $P_1 = p_1 - p_4, P_2 = p_2 - p_1, P_3 = p_3 - p_2, P_4 = p_4 - p_3$ and the diagonals $D_1 = p_1 - p_3, D_2 = p_2 - p_4$. The midpoints of the diagonals are $m_1 = \frac{1}{2}(p_1 - p_3)$ and $m_2 = \frac{1}{2}(p_2 - p_4)$, and the distance between them is $d = |m_2 - m_1| = \frac{1}{2}|p_1 - p_2 - p_3 + p_4|$. The squared length of the sides can be written as $|P_i|^2 = \sum_{j=1}^{3}(p_{i,j} - p_{i-1,j})^2$ where $i = 1, 2, 3, 4$ is cyclic, that is, $i-1 = 4$ for $i = 1$, and j indicates the x, y, and z coordinate of the point respectively. Using the same notation, $|D_i|^2 = \sum_{j=1}^{3}(p_{i,j} - p_{i-2,j})^2$. Then $|P_1|^2 + |P_2|^2 + |P_3|^2 + |P_3|^2 - (|D_1|^2 + |D_2|^2) = \sum_{i=1}^{4}|P_i|^2 - \sum_{i=1}^{2}|D_i|^2 = \sum_{i=1}^{4}\sum_{j=1}^{3}(p_{i,j} - p_{i-1,j})^2 - \sum_{i=1}^{2}\sum_{j=1}^{3}(p_{i,j} - p_{i-2,j})^2$. Exchanging the

order of summation and combining the sums, we get $|P_1|^2 + |P_2|^2 + |P_3|^2 + |P_3|^2 - (|D_1|^2 + |D_2|^2) = \sum_{j=1}^{3} \sum_{i=1}^{4} (p_{i,j} - p_{i-1,j})^2 - \sum_{i=1}^{2} (p_{i,j} - p_{i-2,j})^2 = \sum_{j=1}^{3} \sum_{i=1}^{4} p_{i,j}^2 - 2p_{i,j}p_{i-1,j} + p_{i-1,j}^2 - \sum_{i=1}^{2} p_{i,j}^2 - 2p_{i,j}p_{i-2,j} + p_{i-2,j}^2$. We can see that there is no need to keep track of the individual vector components, because the indices j do not change from term to term. So we write $(|P_1|^2 + |P_2|^2 + |P_3|^2 + |P_3|^2 - (|D_1|^2 + |D_2|^2))_j = L_j = \sum_{i=1}^{4} p_i^2 - 2p_i p_{i-1} + p_{i-1}^2 - \sum_{i=1}^{2} p_i^2 - 2p_i p_{i-2} + p_{i-2}^2$ understanding that this is the same sum for each component $j = 1, 2, 3$ of the vector. Breaking up the first sum and combining, we get

$$\begin{aligned}
L_j &= \sum_{i=1}^{2} p_i^2 - 2p_i p_{i-1} + p_{i-1}^2 - p_i^2 + 2p_i p_{i-2} - p_{i-2}^2 + \sum_{i=3}^{4} p_i^2 - 2p_i p_{i-1} + p_{i-1}^2 \\
&= \sum_{i=1}^{2} p_i^2 - 2p_i p_{i-1} + p_{i-1}^2 - p_i^2 + 2p_i p_{i-2} - p_{i-2}^2 + \sum_{i=1}^{2} p_{i+2}^2 - 2p_{i+2} p_{i+1} + p_{i+1}^2 \\
&= \sum_{i=1}^{2} p_i^2 - 2p_i p_{i-1} + p_{i-1}^2 - p_i^2 + 2p_i p_{i-2} - p_{i-2}^2 + p_{i+2}^2 - 2p_{i+2} p_{i+1} + p_{i+1}^2 \\
&= \sum_{i=1}^{2} -2p_i p_{i-1} + p_{i-1}^2 + 2p_i p_{i-2} - p_{i-2}^2 + p_{i+2}^2 - 2p_{i+2} p_{i+1} + p_{i+1}^2 \\
&= -2p_1 p_4 - 2p_2 p_1 + p_4^2 + p_1^2 + 2p_1 p_3 + 2p_2 p_4 - p_3^2 - p_4^2 + p_3^2 + p_4^2 - 2p_3 p_2 - 2p_4 p_3 + p_2^2 + p_3^2 \\
&= p_1^2 + p_2^2 + p_3^2 + p_4^2 + 2p_1 p_3 + 2p_2 p_4 - 2p_1 p_4 - 2p_2 p_1 - 2p_3 p_2 - 2p_4 p_3
\end{aligned}$$

Now compute the j component of the square of the distance between the midpoints of the diagonals $d_j = (\frac{1}{2}(p_1 - p_3 - (p_2 - p_4)))^2$, where each p_i is really $p_{i,j}$ for $j = 1, 2, 3$, multiply by 4 and see that these are identical.

1.10 Equations of a Plane

1. Find the unit vectors normal to the planes

   (a) $2x + y + 2z = 8$
   (b) $4x - 4z = 0$
   (c) $-y + 6z = 0$
   (d) $x = 5$
   (e) $y = z + 2$
   (f) $x = y$

   Solution: You can read the components of the normal vector **n** right off the equation for a plane: its components are the coefficients of x, y and z.

   (a) $\pm \mathbf{n} = 1/3(2\mathbf{i} + \mathbf{j} + 2\mathbf{k})$
   (b) $\pm \mathbf{n} = \frac{\sqrt{2}}{8}(4\mathbf{i} - 4\mathbf{k})$
   (c) $\pm \mathbf{n} = \frac{\sqrt{37}}{37}(-\mathbf{j} + 6\mathbf{k})$
   (d) $\pm \mathbf{n} = \mathbf{i}$
   (e) $\pm \mathbf{n} = \frac{\sqrt{2}}{2}(\mathbf{j} - \mathbf{k})$
   (f) $\pm \mathbf{n} = \frac{\sqrt{2}}{2}(\mathbf{i} - \mathbf{j})$

3. Find an equation of the plane perpendicular to **D** and through **P**, where

   $$\mathbf{D} = 10\mathbf{i} - 10\mathbf{j} + 5\mathbf{k}$$

   and **P** is $(1, 1, -3)$.
   Solution: The coefficients of x, y and z are the components of **D** and the scalar product of **P** and **D** give the right hand side, so we obtain $10x - 10y + 5z = -15$ or $2x - 2y + z = -3$.

5. Is it possible to find a plane perpendicular to both **i** and **j**?
   Solution: Not in three dimensions, but they are perpendicular to a line in the direction of **k**. If we move to four dimensions and postulate a fourth unit vector **l** in the fourth coordinate direction then **i** and **j** span a plane perpendicular to that spanned by **k** and **l**.

1.10. EQUATIONS OF A PLANE

7. Find the distances between the pairs of planes
   (a) $x + 2y + 3z = 5$ and $x + 2y + 3z = 19$
   (b) $x + y = 4$ and $x + y = 10$
   (c) $x = 5$ and $x = 7$ (no calculations needed here)

   Solution:

   (a) We can use the formula given in example 1.24 for the distance between parallel planes $d = \frac{|ax_1+by_1+cz_1-d|}{\sqrt{a^2+b^2+c^2}}$, where (x_1, y_1, z_1) is a point in one plane and $ax+by+cz = d$ is the formula for the other plane to find $x + 2y + 3z = 5$ and $x + 2y + 3z = 19$ are $d = \frac{|1(14)+2(1)+3(1)-5|}{\sqrt{1^2+2^2+3^2}} = \sqrt{14}$ apart.

   (b) Using the same method as above, we find $x + y = 4$ and $x + y = 10$ are a distance $d = \frac{|1(2)+1(2)+0(0)-10|}{\sqrt{1^2+1^2+0^2}} = 3\sqrt{2}$ apart.

   (c) $x = 5$ and $x = 7$ are a distance 2 apart.

9. By vector methods show that the line $x = y = 1/3(z + 2)$ is parallel to the plane $2x - 8y + 2z = 5$.
   Solution: We can prove this by showing the scalar product of a vector in the direction of the line and the normal to the plane is 0. A vector in the direction of this line is $\mathbf{i}+\mathbf{j}+3\mathbf{k}$ and the normal to the plane is $\mathbf{n} = 2\mathbf{i} - 8\mathbf{j} + 2\mathbf{k}$, whose dot product is zero.

11. Find the angle that the plane OAB makes with the z axis, if A is the point $(1, 3, 2)$ and B is $(2, 1, 1)$.
   Solution: A normal to the plane is $(2\mathbf{i} + \mathbf{j} + \mathbf{k}) \times (\mathbf{i} + 3\mathbf{j} + 2\mathbf{k}) = -\mathbf{i} - 3\mathbf{j} + 5\mathbf{k}$, and the angle between the plane and the z axis will be $90°$ minus the angle between the normal and the z axis. We find $\gamma = 90 - \cos^{-1}\frac{(-\mathbf{i}-3\mathbf{j}+5\mathbf{k})\cdot(\mathbf{k})}{|-\mathbf{i}-3\mathbf{j}+5\mathbf{k}||\mathbf{k}|} = 90 - \cos^{-1}\frac{\sqrt{35}}{7}$

13. By vector methods find the angle between the line $x = y = 2z$ and the plane $x + y + z = 0$.
   Solution: A normal to the plane is $\mathbf{n} = (\mathbf{i}+\mathbf{j}+\mathbf{k})$, and a vector in the direction of the line is $\mathbf{v} = \mathbf{i}+\mathbf{j}+1/2\mathbf{k}$. The angle between the plane and $\mathbf{v}$ will be $90°$ minus the angle between the normal to the plane and $\mathbf{v}$. We find $\theta = 90 - \cos^{-1}\frac{(\mathbf{i}+\mathbf{j}+\mathbf{k})\cdot(\mathbf{i}+\mathbf{j}+1/2\mathbf{k})}{|\mathbf{i}+\mathbf{j}+\mathbf{k}||\mathbf{i}+\mathbf{j}+1/2\mathbf{k}|} = 90 - \cos^{-1}\frac{5}{9}\sqrt{3}$ or $\sin^{-1}\frac{5}{9}\sqrt{3}$

15. Find the equation of a line in the xy plane perpendicular to the vector $3\mathbf{i} - \mathbf{j}$.
   Solution: Any vector in the xy plane has the form $x\mathbf{i}+y\mathbf{j}$, and if its scalar product with $3\mathbf{i}-\mathbf{j}$ is zero, that vector is perpendicular to $3\mathbf{i}-\mathbf{j}$. Taking the scalar product we find $(x\mathbf{i}+y\mathbf{j})\cdot(3\mathbf{i}-\mathbf{j}) = 3x-y = 0$ must hold. Therefore, the most general formula for a line perpendicular to $3\mathbf{i} - \mathbf{j}$ in the xy plane is $\frac{(x-x_0)}{1/3} = \frac{(y-y_0)}{1}$. You can rearrange the formula in the back of the book to put it in this form.

17. Find a line in the xy plane parallel to $3x + 2y = 4$ passing through the point $(3, 1)$.
   Solution: There are many formulas for a line. We will try to get a nice simple one. Write $3x + 2y = 4$ in the form $\frac{x-x_0}{a} = \frac{y-y_0}{b}$ by $3x = -2y + 4$, or $\frac{x}{1/3} = \frac{y-2}{-1/2}$. If we divide this by 6 we get $\frac{x}{2} = \frac{y-2}{-3}$, so a vector in this direction is $2\mathbf{i} - 3\mathbf{j}$, which is arguably easier to work with. A line in this direction containing the point $(3, 1)$ is $\frac{x-3}{2} = \frac{y-1}{-3}$, which can be rearranged to match the answer in the book.

19. We are given two distinct parallel planes and are told the distance between the planes is d. A vector $\mathbf{v}$ is perpendicular to the planes, and its magnitude is $1/d$. The planes intersect the y axis in the points $(0, 1, 0)$ and $(0, 4, 0)$ respectively. What is the y component of $\mathbf{v}$? (There are two possible answers, depending on the two possible directions of $\mathbf{v}$.)
   Solution: Referring to Figure 1.3 we can see that the y axis has a projection onto the perpendicular to the planes that satisfies $d = 3\cos\theta$, so the projection of $\mathbf{v}$, which has length $1/d$, onto the y axis is $1/3 = 1/d\cos\theta$. Had we drawn the vector pointing from the leftmost plane in the picture to the rightmost, the result would be $-1/3$, so the y component is either plus or minus $1/3$.

21. Find the distance from the origin to the plane through $(3, 2, 6)$ that is perpendicular to the z axis.
   Solution: Because the plane is perpendicular to the z axis, it is parallel to the xy plane, and $(0, 0, 6)$ is the point in the plane closest to $(0, 0, 0)$, so the distance of the plane from $(0, 0, 0)$ is 6.

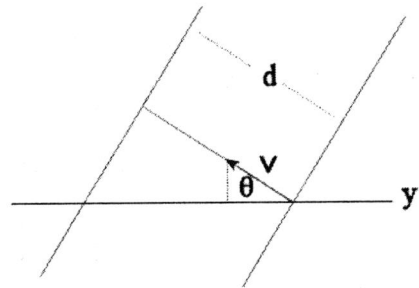

Figure 1.3: The vector **v** projected onto the y axis.

23. A plane has intercepts $(4,0,0)$, $(0,6,0)$ and $(0,0,12)$. Find the equations of another plane through $(6,-2,4)$ that is parallel to this plane.
Solution: Form two vectors $\mathbf{v} = (4,0,0) - (0,6,0) = 4\mathbf{i} - 6\mathbf{j}$ and $\mathbf{u} = (0,0,12) - (0,6,0) = -6\mathbf{j} + 12\mathbf{k}$. Then their vector product $\mathbf{u} \times \mathbf{v} = 72\mathbf{i} + 48\mathbf{j} + 24\mathbf{k}$ is perpendicular to the plane containing $\mathbf{u}$ and $\mathbf{v}$, and is perpendicular to the plane we seek. We can use a scaled down version $3\mathbf{i} + 2\mathbf{j} + \mathbf{k}$ for the normal to make calculation easier; for purposes of finding a formula for a plane, the length of the vector does not matter, only its direction. Dotting and rearranging $(3\mathbf{i}+2\mathbf{j}+\mathbf{k}) \cdot ((x-6)\mathbf{i}+(y+2)\mathbf{j}+(z-4)\mathbf{k}) = 3x - 18 + 2y + 4 + z - 4 = 0$, we get $3x + 2y + z = 18$.

25. What is the distance from the origin to the plane intersecting the x, y, z axes at $x = a$, $y = b$ and $z = c$ respectively?
Solution: We find a vector from some point in the plane to the point in question. This vector can be written as $\mathbf{R}_1 - \mathbf{R}_0$ where $\mathbf{R}_1$ is the position vector to the point and $\mathbf{R}_0$ is the position vector to a point in the plane. The distance from the point to a plane is the projection of this vector on the unit normal to the plane. We already know points in the plane, for instance $(a,0,0)$, so we just need a normal to the plane: $\mathbf{n} = (a\mathbf{i} - b\mathbf{j}) \times (c\mathbf{k} - b\mathbf{j}) = -bc\mathbf{i} - ac\mathbf{j} - ab\mathbf{k}$. The unit normal is $\pm \frac{1}{\sqrt{b^2c^2+a^2c^2+a^2b^2}}(-bc\mathbf{i} - ac\mathbf{j} - ab\mathbf{k})$ and the distance is $d = |\mathbf{n}/|\mathbf{n}| \cdot (\mathbf{R}_1 - \mathbf{R}_0)| = |(\frac{1}{\sqrt{b^2c^2+a^2c^2+a^2b^2}}(-bc\mathbf{i} - ac\mathbf{j} - ab\mathbf{k})) \cdot (-a\mathbf{i})| = \frac{|abc|}{\sqrt{b^2c^2+a^2c^2+a^2b^2}}$.

27. Find the distance between the planes $2x + y + z = 2$ and $2x + y + z = 4$.
Solution: We need only find a point in each plane to form a vector from one point to the other, then project onto the unit normal $\frac{1}{2^2+1^2+1^2}(2\mathbf{i}+\mathbf{j}+\mathbf{k})$. Points in the respective planes are $(0,1,1)$ and $(0,2,2)$ for a vector $\mathbf{j} + \mathbf{k}$ with projection on $\frac{\sqrt{6}}{6}(2\mathbf{i}+\mathbf{j}+\mathbf{k})$ of $\frac{\sqrt{6}}{3}$.

1.11 Orientation

1. If an oriented plane is represented by a vector perpendicular to the area, with magnitude numerically equal to the area, what is the geometrical significance of the components of the vector?
Solution: Each component is the projection of the area onto the coordinate plane.

1.12 Vector Products

1. Find $\mathbf{A} \times \mathbf{B}$, where

   (a) $\mathbf{A} = 3\mathbf{i} - \mathbf{j} + 2\mathbf{k}$, $\mathbf{B} = \mathbf{i} + \mathbf{j} - 4\mathbf{k}$
   (b) $\mathbf{A} = 2\mathbf{i} + \mathbf{j} + 7\mathbf{k}$, $\mathbf{B} = 3\mathbf{i} + \mathbf{j} - \mathbf{k}$
   (c) $\mathbf{A} = \mathbf{j} + 6\mathbf{k}$, $\mathbf{B} = \mathbf{k} + 2\mathbf{j} - \mathbf{i}$
   (d) $\mathbf{A} = \mathbf{i}$, $\mathbf{B} = \mathbf{j}$

1.12. VECTOR PRODUCTS

(e) $\mathbf{B} \times \mathbf{A}$ is known to be $\mathbf{i} - \mathbf{j}$

Solution: Compute the symbolic determinants.

(a) $\begin{vmatrix} \mathbf{i} & \mathbf{j} & \mathbf{k} \\ 3 & -1 & 2 \\ 1 & 1 & -4 \end{vmatrix} = 2\mathbf{i} + 14\mathbf{j} + 4\mathbf{k}$

(b) $\begin{vmatrix} \mathbf{i} & \mathbf{j} & \mathbf{k} \\ 2 & 1 & 7 \\ 3 & 1 & -1 \end{vmatrix} = -8\mathbf{i} + 23\mathbf{j} - \mathbf{k}$

(c) $\begin{vmatrix} \mathbf{i} & \mathbf{j} & \mathbf{k} \\ 0 & 1 & 6 \\ -1 & 2 & 1 \end{vmatrix} = -11\mathbf{i} - 6\mathbf{j} + \mathbf{k}$

(d) $\begin{vmatrix} \mathbf{i} & \mathbf{j} & \mathbf{k} \\ 1 & 0 & 0 \\ 0 & 1 & 0 \end{vmatrix} = \mathbf{k}$

(e) Reversing the order of the terms in the vector product reverses the sign, so $\mathbf{A} \times \mathbf{B} = -\mathbf{i} + \mathbf{k}$

3. Find the area of a triangle with vertices $(1, 1, 2)$, $(2, 3, 5)$ and $(1, 5, 5)$.
   Solution: The area is one half the magnitude of the cross product of two vectors between pairs of points. So take $\mathbf{u} = (2, 3, 5) - (1, 1, 2) = \mathbf{i} + 2\mathbf{j} + 3\mathbf{k}$ and $\mathbf{v} = (2, 3, 5) - (1, 5, 5) = \mathbf{i} - 2\mathbf{j}$. The area of the parallelogram with these vectors as sides is given by the magnitude of the cross product of these two vectors: $b\mathbf{u} \times \mathbf{v} = \begin{vmatrix} \mathbf{i} & \mathbf{j} & \mathbf{k} \\ 1 & 2 & 3 \\ 1 & -2 & 0 \end{vmatrix} = 6\mathbf{i} + 3\mathbf{j} - 4\mathbf{k}$. Now take the magnitude and divide by two to get the area of the triangle $\sqrt{6^2 + 3^2 + (-4)^2}/2 = \sqrt{61}/2$.

5. Find a unit vector perpendicular to both $3\mathbf{i} + \mathbf{j}$ and $2\mathbf{i} - \mathbf{j} - 5\mathbf{k}$.
   Solution: There are two vectors that fit the bill, and they point in opposite directions:
   $$\mathbf{N} = \pm \frac{(3\mathbf{i} + \mathbf{j}) \times (2\mathbf{i} - \mathbf{j} - 5\mathbf{k})}{|(3\mathbf{i} + \mathbf{j}) \times (2\mathbf{i} - \mathbf{j} - 5\mathbf{k})|} = \pm \frac{(-5\mathbf{i} + 15\mathbf{j} - 5\mathbf{k})}{\sqrt{(-5)^2 + 15^2 + (-5)^2}} = \pm \frac{\sqrt{11}}{11}(-\mathbf{i} + 3\mathbf{j} - \mathbf{k}).$$

7. Find equations of a line perpendicular to the lines $x = y = z$ and $x = 2y = 3z$, passing through the origin.
   Solution: The normals are $\mathbf{i} + \mathbf{j} + \mathbf{k}$ and $\mathbf{i} + 1/2\mathbf{j} + 1/3\mathbf{k}$ and perpendicular to these is $(\mathbf{i} + 1/2\mathbf{j} + 1/3\mathbf{k}) \times (\mathbf{i} + \mathbf{j} + \mathbf{k}) = 1/6\mathbf{i} + 2/3\mathbf{j} + 1/2\mathbf{k}$. We can multiply by 6 to get rid of fractions, then the equation of the line is $\frac{x-0}{1} = \frac{y-0}{4} = \frac{z-0}{3}$.

9. By vector methods, determine the equation of the plane determined by the points $(2, 0, 1)$, $(1, 1, 3)$ and $(4, 7, -2)$.
   Solution: Take the vector product of two vectors $\mathbf{u} = (2, 0, 1) - (1, 1, 3) = \mathbf{i} - \mathbf{j} - 2\mathbf{k}$ and $\mathbf{v} = (4, 7, -2) - (1, 1, 3) = 3\mathbf{i} + 6\mathbf{j} - 5\mathbf{k}$ in the plane to get a normal: $\mathbf{u} \times \mathbf{v} = (\mathbf{i} - \mathbf{j} - 2\mathbf{k}) \times (3\mathbf{i} + 6\mathbf{j} - 5\mathbf{k}) = 17\mathbf{i} - \mathbf{j} + 9\mathbf{k}$. Then an equation of the plane can be found by computing $(17\mathbf{i} - \mathbf{j} + 9\mathbf{k}) \cdot ((x-1)\mathbf{i} + (y-1)\mathbf{j} + (z-3)\mathbf{k})$ or $17x - y + 9z = 43$

11. By taking the vector cross product of $(\cos\theta)\mathbf{i} + (\sin\theta)\mathbf{j}$ and $(\cos\psi)\mathbf{i} + (\sin\psi)\mathbf{j}$ and interpreting geometrically, derive a well-known trigonometric identity.
    Solution: We assume without loss of generality that $\psi > \theta$. The area of the parallelogram given by the vectors $(\cos\theta)\mathbf{i} + (\sin\theta)\mathbf{j}$ and $(\cos\psi)\mathbf{i} + (\sin\psi)\mathbf{j}$ is the length 1 of the base $\cos\theta\mathbf{i} + \sin\theta\mathbf{j}$ times the sine of the angle between the vectors $\psi - \theta$. The area of the parallelogram is also given by the cross magnitude of the vector product $\begin{vmatrix} \mathbf{i} & \mathbf{j} & \mathbf{k} \\ \cos\theta & \sin\theta & 0 \\ \cos\psi & \sin\psi & 0 \end{vmatrix} = (\cos\theta\sin\psi - \cos\psi\sin\theta)\mathbf{k}$. Thus we have $\sin(\psi - \theta) = \cos\theta\sin\psi - \cos\psi\sin\theta)$.

13. Find the distance from point $(5, 7, 14)$ to the line passing through $(2, 3, 8)$ and $(3, 6, 12)$.
    Solution: The three points lie in a plane, and we can find the distance from a point to a line through two others if we can find a unit vector perpendicular to the line and in the plane of the points. Then a vector from any point in the line to the point not in the line will have a projection on the unit vector that is the minimum distance from the line to the point (see Figure 1.4). First, find a vector parallel to the line: $\mathbf{w} = (3, 6, 12) - (2, 3, 8) = \mathbf{i} + 3\mathbf{j} + 4\mathbf{k}$. Then an equation of the

line is (using the point $(2,3,8)$) is $\frac{x-2}{1} = \frac{y-3}{3} = \frac{z-8}{4}$. To find a normal to the line, we compute a normal $\mathbf{n}$ to the plane, then take its vector product with the vector parallel to the line. The resulting vector will be perpendicular to both, so it will also be perpendicular to the line and will lie in the plane. We already have one vector in the plane, $\mathbf{w} = \mathbf{i} + 3\mathbf{j} + 4\mathbf{k}$. Another is $\mathbf{v} = (5,7,14) - (3,6,12) = 2\mathbf{i} + \mathbf{j} + 2\mathbf{k}$, which we will also use when we project it on the normal to the line. Their vector product $\mathbf{n} = (\mathbf{i} + 3\mathbf{j} + 4\mathbf{k}) \times (2\mathbf{i} + \mathbf{j} + 2\mathbf{k}) = 2\mathbf{i} + 6\mathbf{j} - 5\mathbf{k}$ is perpendicular to the plane. The vector product $\mathbf{n} \times \mathbf{w} = (2\mathbf{i} + 6\mathbf{j} - 5\mathbf{k}) \times (\mathbf{i} + 3\mathbf{j} + 4\mathbf{k}) = 39\mathbf{i} - 13\mathbf{j}$ normalizes to $\mathbf{N} = \frac{\sqrt{10}}{130}(39\mathbf{i} - 13\mathbf{j})$. Taking the projection gives us the distance of the point from the line: $\mathbf{v} = (\frac{\sqrt{10}}{130}(39\mathbf{i} - 13\mathbf{j})) \cdot (2\mathbf{i} + \mathbf{j} + 2\mathbf{k}) = \frac{\sqrt{10}}{2}$.

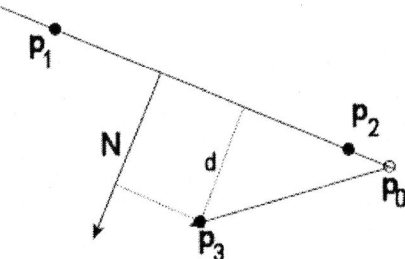

Figure 1.4: The projection of a vector from a point in the line to a point off the line onto a unit normal to the line gives the distance from the line to the point.

15. Write the scalar equations of the line parallel to the intersection of the planes $3x + y + z = 5$, $x - 2y + 3z = 1$, and passing through the point $(4, 2, 1)$.
    Solution: A vector in the direction of the line of intersection of the planes is found by taking the vector product of the normals to the planes: $\mathbf{v} = (3\mathbf{i} + \mathbf{j} + \mathbf{k}) \times (\mathbf{i} - 2\mathbf{j} + 3\mathbf{k}) = 5\mathbf{i} - 8\mathbf{j} - 7\mathbf{k}$. A line in this direction passing through $(4,2,1)$ is $\frac{x-4}{5} = \frac{y-2}{-8} = \frac{z-1}{-7}$.

17. Write an expression for a vector five units long, parallel to the plane $3x + 4y + 5z = 10$ and perpendicular to the vector $\mathbf{i} + 2\mathbf{j} + 2\mathbf{k}$.
    Solution: A vector $\mathbf{v}$ parallel to the plane and perpendicular to $\mathbf{i} + 2\mathbf{j} + 2\mathbf{k}$ is the vector product of the normal to the plane $\mathbf{n} = 3\mathbf{i} + 4\mathbf{j} + 5\mathbf{k}$ and $\mathbf{i} + 2\mathbf{j} + 2\mathbf{k}$: $\mathbf{u} = (3\mathbf{i} + 4\mathbf{j} + 5\mathbf{k}) \times (\mathbf{i} + 2\mathbf{j} + 2\mathbf{k}) = -2\mathbf{i} - \mathbf{j} + 2\mathbf{k}$. The vector we want is five units long, so find the unit vector $\mathbf{U} = 1/3(-2\mathbf{i} - \mathbf{j} + 2\mathbf{k})$ and multiply by 5 to get $\pm 5/3(-2\mathbf{i} - \mathbf{j} + 2\mathbf{k})$, where the $\pm$ indicates that the vector in either direction fulfills the qualifications.

19. Given that $\mathbf{A} \cdot \mathbf{B} = 0$ and $\mathbf{A} \times \mathbf{B} = 0$, what can you conclude about the vectors $\mathbf{A}$ and $\mathbf{B}$?
    Solution: If the vectors were nonzero, $\mathbf{A} \cdot \mathbf{B} = 0$ indicates that they are orthogonal (perpendicular). If they are nonzero and perpendicular, their vector product would be nonzero, so one or both must be zero.

21. Express $2\mathbf{i} - \mathbf{j} + 3\mathbf{k}$ as the sum of a vector parallel, plus a vector perpendicular, to $2\mathbf{i} + 4\mathbf{j} - 2\mathbf{k}$.
    Solution: Using the formula $\mathbf{B}_{\parallel} = \frac{\mathbf{B} \cdot \mathbf{A}}{\mathbf{A} \cdot \mathbf{A}} \mathbf{A}$, and with $\mathbf{A} = 2\mathbf{i} + 4\mathbf{j} - 2\mathbf{k}$ and $\mathbf{B} = 2\mathbf{i} - \mathbf{j} + 3\mathbf{k}$, we find the parallel component is $\mathbf{B}_{\parallel} = -1/2\mathbf{i} - \mathbf{j} + 1/2\mathbf{k}$. The perpendicular component is found by subtracting the parallel component from the original vector $\mathbf{B}_{\perp} = (2\mathbf{i} - \mathbf{j} + 3\mathbf{k}) - (-1/2\mathbf{i} - \mathbf{j} + 1/2\mathbf{k}) = 5/2\mathbf{i} + 5/2\mathbf{k}$.

23. (a) Do the lines $x/3 = y/2 = z/2$ and $x/5 = y/3 = (z-4)/2$ intersect?
    (b) Find equations for a line perpendicular to both of these lines.
    (c) What is the distance between these lines?
    Solution:

    (a) Solving the pairs of equalities $x/3 = y/2$ and $x/5 = y/3$ simultaneously shows that $x = 3/2y$ and $x = 5/3y$, so they cannot have a point in common unless it is $(0,0,z)$. Because we must also have $y/2 = z/2$ and $y/3 = (z-4)/2$, $y = z = 0$ and $z = 4$ must hold, so the lines have no point in common.
    (b) A line perpendicular to both of these lines can be found by computing the vector product of the vectors $\mathbf{u} = 3\mathbf{i} + 2\mathbf{j} + 2\mathbf{k}$ and $\mathbf{w} = 5\mathbf{i} + 3\mathbf{j} + 2\mathbf{k}$ in the directions of the lines: $\mathbf{v} = \mathbf{w} \times \mathbf{u} = (5\mathbf{i} + 3\mathbf{j} + 2\mathbf{k}) \times (3\mathbf{i} + 2\mathbf{j} + 2\mathbf{k}) = 2\mathbf{i} - 4\mathbf{j} + \mathbf{k}$. Then a line in this direction is of the form $\frac{x-x_0}{2} = \frac{y-y_0}{-4} = \frac{z-z_0}{1}$.

1.12. VECTOR PRODUCTS

(c) The distance between these lines is found by finding a unit vector perpendicular to both and projecting onto a vector connecting any point in one line to any point in the other. A unit vector perpendicular to both is $\mathbf{n} = \frac{\sqrt{21}}{21}(2\mathbf{i} - 4\mathbf{j} + \mathbf{k})$, and a vector between the two lines is $\mathbf{v} = (0,0,4) - (0,0,0) = 4\mathbf{k}$. The projection is $\frac{2\sqrt{21}}{21}$.

25. Supply the missing details of the proof of the distributive law for vector products. (Use similar triangles.)

Solution: Writing $\mathbf{A} \times (\mathbf{B} + \mathbf{C}) = \begin{vmatrix} \mathbf{i} & \mathbf{j} & \mathbf{k} \\ A_1 & A_2 & A_3 \\ B_1+C_1 & B_2+C_2 & B_3+C_3 \end{vmatrix} = (A_2(B_3+C_3) - A_3(B_2+C_2))\mathbf{i} + (A_3(B_1+C_1) - A_1(B_3+C_3))\mathbf{j} + (A_1(B_2+C_2) - A_2(B_1+C_1))\mathbf{k} = (A_2 B_3 - A_3 B_2)\mathbf{i} + (A_3 B_1 - A_1 B_3)\mathbf{j} + (A_1 B_2 - A_2 B_1)\mathbf{k} + (A_2 C_3 - A_3 C_2)\mathbf{i} + (A_3 C_1 - A_1 C_3)\mathbf{j} + (A_1 C_2 - A_2 C_1)\mathbf{k} = \mathbf{A} \times \mathbf{B} + \mathbf{A} \times \mathbf{C}$.

27. Given two nonintersecting lines $x/2 = y = (z-1)/3$ and $x/3 = y = z$, find points P and Q, one on each line, such that PQ is perpendicular to both lines.
Solution: Find two planes, each containing one of the lines as well as the normal vector, and take their intersection. The intersection will be the line that connects P and Q. It is then a simple matter of solving pairs of equations for this line and each of the original lines to find P and Q. Vectors $\mathbf{u} = 2\mathbf{i} + \mathbf{j} + 3\mathbf{k}$ and $\mathbf{v} = 3\mathbf{i} + \mathbf{j} + \mathbf{k}$ are in the directions of the lines, so a normal to them both is $\mathbf{n} = \mathbf{u} \times \mathbf{v} = -2\mathbf{i} + 7\mathbf{j} - \mathbf{k}$. Vectors normal to $\mathbf{u}$ and $\mathbf{n}$ and to $\mathbf{v}$ and $\mathbf{n}$ define the planes whose intersection we seek. These are $\mathbf{A} = \mathbf{u} \times \mathbf{n} = -22\mathbf{i} - 4\mathbf{j} + 16\mathbf{k}$ and $\mathbf{B} = \mathbf{v} \times \mathbf{n} = -8\mathbf{i} + \mathbf{j} + 23\mathbf{k}$, and these define planes $(-22\mathbf{i} - 4\mathbf{j} + 16\mathbf{k}) \cdot ((x - x_0)\mathbf{i} + (y - y_0)\mathbf{j} + (z - z_0)\mathbf{k}) = 0$ and $(-8\mathbf{i} + \mathbf{j} + 23\mathbf{k}) \cdot ((x - x_0)\mathbf{i} + (y - y_0)\mathbf{j} + (z - z_0)\mathbf{k}) = 0$. Inserting points $(x_0, y_0, z_0) = (0,0,1)$ and $(x_0, y_0, z_0) = (0,0,0)$ from the first and second line respectively, we get $-22x - 4y + 16z = 16$ and $-8x + y + 23z = 0$. Solving these simultaneously, we find the relation $x - 2z = -8/27$. Solving this simultaneously with the equations for the lines yields $P = (-23/27, -23/54, -5/18)$ and $Q = (-8/9, -8/27, -8/27)$.

29. *Ampere's force* is the magnetic force that one moving charged particle exerts on another moving charged particle. Suppose particle 1 has charge q_1 and velocity $\mathbf{v}_1$, and is located at $\mathbf{R}_1$; q_2, $\mathbf{v}_2$ and $\mathbf{R}_2$ are the corresponding parameters for particle 2. Particle 2 produces a "magnetic flux density" vector $\mathbf{B}$ at $\mathbf{R}_1$ that is proportional to q_2 and $|\mathbf{v}_2|$ and inversely proportional to the square of the distance between q_2 to q_1. The force on particle 1 is proportional to q_1, $|\mathbf{B}|$, and $|\mathbf{v}_1|$, and is directed perpendicular to $\mathbf{v}_1$ and to $\mathbf{B}$. Show that the formula

$$\text{Force on particle 1} = k\, q_1 q_2\, \mathbf{v}_1 \times \left\{ \mathbf{v}_2 \times \frac{\mathbf{R}_1 - \mathbf{R}_2}{|\mathbf{R}_1 - \mathbf{R}_2|^3} \right\}$$

has all these properties.
Solution: From the description of the problem we have

$$|\mathbf{B}| = \frac{k_1 q_2 |\mathbf{v}_2|}{|\mathbf{R}_1 - \mathbf{R}_2|^2}$$

and $\mathbf{B} = |\mathbf{B}| \left(\frac{\mathbf{v}_2}{|\mathbf{v}_2|} \times \frac{\mathbf{R}_1 - \mathbf{R}_2}{|\mathbf{R}_1 - \mathbf{R}_2|} \right)$. In addition, we are given $\mathbf{F}_1 = k_2 q_1 |\mathbf{B}||\mathbf{v}_1| \left(\frac{\mathbf{v}_1}{|\mathbf{v}_1|} \times \frac{\mathbf{B}}{|\mathbf{B}|} \right) = k_2 q_1 (\mathbf{v}_1 \times \mathbf{B})$. Putting it all together, we have

$$\mathbf{F}_1 = k_2 q_1 \mathbf{v}_1 \times \left(\frac{k_1 q_2 |\mathbf{v}_2|}{|\mathbf{R}_1 - \mathbf{R}_2|^2} \frac{\mathbf{v}_2}{|\mathbf{v}_2|} \times \frac{\mathbf{R}_1 - \mathbf{R}_2}{|\mathbf{R}_1 - \mathbf{R}_2|} \right) = k_1 k_2 q_1 q_2 \mathbf{v}_1 \times \left(\mathbf{v}_2 \times \frac{\mathbf{R}_1 - \mathbf{R}_2}{|\mathbf{R}_1 - \mathbf{R}_2|^3} \right)$$

31. What is the force on particle 2 in the situations depicted in figure 1.37? Is it equal and opposite to the force on particle 1?
Solution: From the form of the equation giving the force of one particle on the other $\mathbf{F}_{12} = k_2 q_1 q_2 \mathbf{v}_1 \times \left(\mathbf{v}_2 \times \frac{\mathbf{R}_1 - \mathbf{R}_2}{|\mathbf{R}_1 - \mathbf{R}_2|^3} \right)$ and $\mathbf{F}_{21} = k_2 q_2 q_1 \mathbf{v}_2 \times \left(\mathbf{v}_1 \times \frac{\mathbf{R}_1 - \mathbf{R}_2}{|\mathbf{R}_1 - \mathbf{R}_2|^3} \right)$ it is apparent that the asymmetry due to the order of the cross products will not matter only in the case that $\mathbf{v}_1 = \mathbf{v}_2$, so in general the force will not be equal and opposite. To show this more clearly, let $\mathbf{v}_3 = \frac{\mathbf{R}_1 - \mathbf{R}_2}{|\mathbf{R}_1 - \mathbf{R}_2|}$, then take the sum $\mathbf{v}_1 \times (\mathbf{v}_2 \times \mathbf{v}_3) + \mathbf{v}_2 \times (\mathbf{v}_1 \times \mathbf{v}_3)$. This should be identically zero if the forces are, in general, equal and opposite.

1.13 Triple Scalar Products

1. Find the triple scalar product $[\mathbf{A}, \mathbf{B}, \mathbf{C}]$ given that

   (a) $\mathbf{A} = 2\mathbf{i}$, $\mathbf{B} = 3\mathbf{j}$, $\mathbf{C} = 5\mathbf{k}$
   (b) $\mathbf{A} = \mathbf{i} + \mathbf{j} + \mathbf{k}$, $\mathbf{B} = 3\mathbf{i} + \mathbf{j}$, $\mathbf{C} = 5\mathbf{k} - \mathbf{j}$
   (c) $\mathbf{A} = 2\mathbf{i} - \mathbf{j} + \mathbf{k}$, $\mathbf{B} = \mathbf{i} + \mathbf{j} + \mathbf{k}$, $\mathbf{C} = 2\mathbf{i} + 3\mathbf{k}$
   (d) $\mathbf{A} = \mathbf{k}$, $\mathbf{B} = \mathbf{i}$, $\mathbf{C} = \mathbf{j}$

   Solution: The triple scalar product can be calculated as the determinant of a 3×3 matrix. This is completely equivalent to taking the vector product of two vectors first then dotting the third vector into the result.

   (a) $\begin{vmatrix} 2 & 0 & 0 \\ 0 & 3 & 0 \\ 0 & 0 & 5 \end{vmatrix} = 30$

   (b) $\begin{vmatrix} 1 & 1 & 1 \\ 3 & 1 & 0 \\ 0 & -1 & 5 \end{vmatrix} = 5 - 15 - 3 = -13$

   (c) $\begin{vmatrix} 2 & -1 & 1 \\ 1 & 1 & 1 \\ 2 & 0 & 3 \end{vmatrix} = 6 + 1 - 2 = 5$

   (d) $\begin{vmatrix} 0 & 0 & 1 \\ 1 & 0 & 0 \\ 0 & 1 & 0 \end{vmatrix} = 1$

3. Find the volume of the parallelepiped with coterminal edges AB, AC, and AD, where $A = (3, 2, 1)$, $B = (4, 2, 1)$, $C = (0, 1, 4)$, and $D = (0, 0, 7)$.
   Solution: The volume of a parallelepiped can be computed by taking the absolute value of the triple scalar product. The three vectors anchored at A are $\mathbf{u} = B - A = (4, 2, 1) - (3, 2, 1) = \mathbf{i}$, $\mathbf{v} = C - A = (0, 1, 4) - (3, 2, 1) = -3\mathbf{i} - \mathbf{j} + 3\mathbf{k}$, and $\mathbf{w} = D - A = (0, 0, 7) - (3, 2, 1) = -3\mathbf{i} - 2\mathbf{j} + 6\mathbf{k}$.
   We compute their triple scalar product $\begin{vmatrix} 1 & 0 & 0 \\ -3 & -1 & 3 \\ -3 & -2 & 6 \end{vmatrix} = 0$. This indicates that these three vectors are all in the same plane. Note that the angle θ in the expression for the triple scalar product $[\mathbf{B}, \mathbf{C}, \mathbf{A}] = [\mathbf{A}, \mathbf{B}, \mathbf{C}] = [\mathbf{C}, \mathbf{A}, \mathbf{B}] = \mathbf{C} \cdot (\mathbf{A} \times \mathbf{B}) = |\mathbf{C}||\mathbf{A} \times \mathbf{B}|\cos\theta$ must be 0 in this case: the vector product of two vectors is perpendicular to the plane in which those vectors lie.

5. Find the area of the parallelogram in the plane with vertices at $(0, 0)$, $(1, 1)$, $(3, 4)$, $(4, 5)$. (*Hint:* Convert this to a three dimensional problem, finding the volume of the parallelepiped with this parallelogram as base, taking the third edge to be of unit length along the z axis.
   Solution: We can calculate this as the author hints, calculating the triple scalar product $\mathbf{A} \cdot (\mathbf{B} \times \mathbf{C}) = 1$ where $\mathbf{A} = \mathbf{k}$, $\mathbf{B} = (1, 1, 0) - (0, 0, 0) = \mathbf{i} + \mathbf{j}$, and $\mathbf{C} = (3, 4, 0) - (0, 0, 0) = 3\mathbf{i} + 4\mathbf{j}$. By examining the form of the determinant from this method, we can see that for any pair of two dimensional vectors, we can calculate the area of the parallelogram they span just by taking the absolute value of the determinant of the 2×2 matrix with these vectors as rows or columns. The calculation goes as follows: $\begin{vmatrix} 1 & 1 \\ 3 & 4 \end{vmatrix} = 1$ or $\begin{vmatrix} 1 & 3 \\ 1 & 4 \end{vmatrix} = 1$.

7. Find the equation of the plane passing through $(3, 4, -1)$ parallel to the vectors $\mathbf{A} = 2\mathbf{i} + \mathbf{j} + \mathbf{k}$ and $\mathbf{B} = \mathbf{i} - 3\mathbf{k}$.
   Solution: Recall that the formula for a plane is constructed by defining the set of points for which the scalar product of a vector $\mathbf{n} = a\mathbf{i} + b\mathbf{j} + c\mathbf{k}$ and a vector $\mathbf{v} = (x - x_0)\mathbf{i} + (y - y_0)\mathbf{j} + (z - z_0)\mathbf{k}$ that is variable but thought of as anchored in space at $x_0\mathbf{i} + y_0\mathbf{j} + z_0\mathbf{k}$. The set of points for which $\mathbf{n} \cdot \mathbf{v} = (a\mathbf{i} + b\mathbf{j} + c\mathbf{k}) \cdot ((x - x_0)\mathbf{i} + (y - y_0)\mathbf{j} + (z - z_0)\mathbf{k}) = 0$ is the set of points in the plane normal to $\mathbf{n} = a\mathbf{i} + b\mathbf{j} + c\mathbf{k}$. When we multiply this out, we get the formula for a plane $ax + by + cz = d$ where $d = ax_0 + by_0 + cz_0$. We have the point in the plane (x_0, y_0, z_0), and can get a vector perpendicular to the vectors $\mathbf{A}$ and $\mathbf{B}$ by taking their vector product $\mathbf{A} \times \mathbf{B} = -3\mathbf{i} + 7\mathbf{j} - \mathbf{k}$, so the formula for the plane is $-3x + 7y - z = 20$.

9. Given the points $P_1(2, -1, 4)$, $P_2(-1, 0, 3)$, $P_3(4, 3, 1)$, and $P_4(3, -5, 0)$, determine

1.13. TRIPLE SCALAR PRODUCTS

(a) the volume of the tetrahedron $P_1P_2P_3P_4$.

(b) the equation of the plane containing the points P_1, P_2, and P_3.

(c) the cosine of the angle between the line segments P_1P_2 and P_1P_3.

Solution: It is useful to calculate the vectors $\mathbf{u} = P_1 - P_2 = 3\mathbf{i} - \mathbf{j} + \mathbf{k}$, $\mathbf{v} = P_3 - P_2 = 5\mathbf{i} + 3\mathbf{j} - 2\mathbf{k}$ and $\mathbf{w} = P_4 - P_2 = 4\mathbf{i} - 5\mathbf{j} - 3\mathbf{k}$.

(a) The volume of the tetrahedron $P_1P_2P_3P_4$ is $1/6$ the volume of the parallelepiped given by the vectors $\mathbf{u}, \mathbf{v}, \mathbf{w}$ or $|[\mathbf{uvw}]|/6 = 101/6$.

(b) We have already formed the vector joining the points P_1 and P_3 to P_2: $\mathbf{u} = P_1 - P_2 = 3\mathbf{i} - \mathbf{j} + \mathbf{k}$, $\mathbf{v} = P_3 - P_2 = 5\mathbf{i} + 3\mathbf{j} - 2\mathbf{k}$. The equation for a plane containing these can be determined by choosing one point $P_2 = (x_0, y_0, z_0)$ as a base point and forming the vector product $\mathbf{u} \times \mathbf{v} = (3\mathbf{i} - \mathbf{j} + \mathbf{k}) \times (5\mathbf{i} + 3\mathbf{j} - 2\mathbf{k}) = -\mathbf{i} + 11\mathbf{j} + 14\mathbf{k}$, which will be our normal vector. Then an equation of our plane is $-x + 11y + 14z = 43$, which can be manipulated to match the answer in the book.

(c) The cosine of the angle between the line segments P_1P_2 and P_1P_3 is $\cos\theta = \frac{(P_2-P_1)\cdot(P_3-P_1)}{|P_2-P_1||P_3-P_1|} = \frac{(-3\mathbf{i}+\mathbf{j}-\mathbf{k})\cdot(2\mathbf{i}+4\mathbf{j}-3\mathbf{k})}{|-3\mathbf{i}+\mathbf{j}-\mathbf{k}||2\mathbf{i}+4\mathbf{j}-3\mathbf{k}|} = \frac{1}{\sqrt{319}}$.

11. Find the altitude of a parallelepiped determined by $\mathbf{A}$, $\mathbf{B}$, and $\mathbf{C}$, if the base is taken to be the parallelogram determined by $\mathbf{A}$ and $\mathbf{B}$, and if

$$\mathbf{A} = \mathbf{i} + \mathbf{j} + \mathbf{k}, \qquad \mathbf{B} = 2\mathbf{i} + 4\mathbf{j} - \mathbf{k}, \qquad \mathbf{C} = \mathbf{i} + \mathbf{j} + 3\mathbf{k}$$

(*Hint*: Think of the geometrical interpretation of $[\mathbf{A}, \mathbf{B}, \mathbf{C}]/\|\mathbf{A} \times \mathbf{B}\|$)
Solution: The volume of the parallelepiped is given both by $V = bh$, where the area of the base can be computed as the magnitude of the cross product of the vectors forming the parallelogram, or as the absolute value of the triple scalar product. We can compute the altitude h by dividing the volume of the parallelepiped by the base area: $h = V/b = |[\mathbf{A},\mathbf{B},\mathbf{C}]|/|\mathbf{A}\times\mathbf{B}| = \frac{2}{19}\sqrt{38}$.

13. Given the four points specified in exercise 12 ($A = (1, 3, -2)$, $B = (3, 5, -3)$, $C = (-5, 9, -5)$ and $D = (4, -1, 10)$, determine

(a) the area of the triangle OAB.

(b) the volume of the tetrahedron $OABC$.

(c) the angle CAD.

Solution:

(a) The area of the triangle OAB is one half the area of the parallelogram given by the vectors $\mathbf{A}$ and $\mathbf{B}$ or $\frac{1}{2}|\mathbf{A}\times\mathbf{B}| = \frac{1}{2}|(\mathbf{i}+3\mathbf{j}-2\mathbf{k})\times(3\mathbf{i}+5\mathbf{j}-3\mathbf{k})| = \frac{1}{2}\sqrt{26}$.

(b) The volume of the tetrahedron $OABC$ is $1/6$ the absolute value of the triple scalar product of $\mathbf{A}, \mathbf{B}$ and $\mathbf{C}$, or $V = 1/6|[\mathbf{A},\mathbf{B},\mathbf{C}]| = 1/6|((\mathbf{i}+3\mathbf{j}-2\mathbf{k})\times(3\mathbf{i}+5\mathbf{j}-3\mathbf{k}))\cdot(-5\mathbf{i}+9\mathbf{j}-5\mathbf{k})| = 2$.

(c) The angle CAD is $\cos^{-1}\frac{(C-A)\cdot(B-A)}{|(C-A)||(B-A)|} = \cos^{-1}\frac{(-6\mathbf{i}+6\mathbf{j}-3\mathbf{k})\cdot(2\mathbf{i}+2\mathbf{j}-\mathbf{k})}{|(-6\mathbf{i}+6\mathbf{j}-3\mathbf{k})||(2\mathbf{i}+2\mathbf{j}-\mathbf{k})|} = \cos^{-1} 1/9$.

15. What can you conclude about nonzero vectors $\mathbf{A}, \mathbf{B}, \mathbf{C}$, and $\mathbf{D}$, given that $|(\mathbf{A}\times\mathbf{B})\cdot\mathbf{C}| + |(\mathbf{B}\times\mathbf{C})\cdot\mathbf{D}| = 0$?
Solution: If any of the vectors in either group is zero, the triple scalar product of that group will be zero. If the vectors are all assumed nonzero, then we know that because both quantities are positive, both triple scalar products are zero. Therefore $\mathbf{C}$ is in the plane spanned by $\mathbf{A}$ and $\mathbf{B}$, and also in the plane spanned by $\mathbf{B}$ and $\mathbf{D}$. Because any two vectors determine a plane, all four vectors are coplanar.

17. Show that an arbitrary vector $\mathbf{V}$ can be expressed in terms of any three non-coplanar vectors $\mathbf{A}, \mathbf{B}$, and $\mathbf{C}$, according to

$$\mathbf{V} = \frac{[\mathbf{V},\mathbf{B},\mathbf{C}]}{[\mathbf{A},\mathbf{B},\mathbf{C}]}\mathbf{A} + \frac{[\mathbf{V},\mathbf{C},\mathbf{A}]}{[\mathbf{A},\mathbf{B},\mathbf{C}]}\mathbf{B} + \frac{[\mathbf{V},\mathbf{A},\mathbf{B}]}{[\mathbf{A},\mathbf{B},\mathbf{C}]}\mathbf{C}$$

(*Hint*: We know that **V** can be expressed as $a\mathbf{A}+b\mathbf{B}+c\mathbf{C}$; to find a, take the scalar product of **V** with $\mathbf{B} \times \mathbf{C}$.)
Solution: We know there is some linear combination of the three vectors **A**, **B** and **C** giving **V** as $\mathbf{V} = a\mathbf{A} + b\mathbf{B} + c\mathbf{C}$, so following the hint we dot both sides with $\mathbf{B} \times \mathbf{C}$ to get $\mathbf{V} \cdot \mathbf{B} \times \mathbf{C} = (a\mathbf{A} + b\mathbf{B} + c\mathbf{C}) \cdot (\mathbf{B} \times \mathbf{C}) = a\mathbf{A}\mathbf{B} \times \mathbf{C}$ where we have used the fact that $\mathbf{Q} \cdot \mathbf{Q}\mathbf{P} = 0$. Now we can solve for a: $\mathbf{V} \times \mathbf{B} \times \mathbf{C} = a\mathbf{A}\mathbf{B} \times \mathbf{C} \Rightarrow a = \frac{[\mathbf{V},\mathbf{B},\mathbf{C}]}{[\mathbf{A},\mathbf{B},\mathbf{C}]}$. Similarly, we multiply by $\mathbf{B} \times \mathbf{A}$ and $\mathbf{C} \times \mathbf{A}$ to solve for b and c.

19. Construct another proof of the distributive law for the vector product, based on the interchange of $\times$ and $\cdot$ (see exercise 18) and the distributivity of the scalar product. (*Hint:* Derive the identity

$$\mathbf{D} \cdot \mathbf{A} \times (\mathbf{B} + \mathbf{C}) = \mathbf{D} \cdot \mathbf{A} \times \mathbf{B} + \mathbf{D} \cdot \mathbf{A} \times b\mathbf{C}$$

and then let **D** be **i**, **j**, and **k**, in turn.)
Solution: $\mathbf{D} \cdot \mathbf{A} \times (\mathbf{B} \times \mathbf{C}) = [\mathbf{D},\mathbf{A},(\mathbf{B}+\mathbf{C})] = (\mathbf{B}+\mathbf{C}) \cdot (\mathbf{D} \times \mathbf{A})$. By the distributive law for the scalar product $(\mathbf{B}+\mathbf{C}) \cdot (\mathbf{D} \times \mathbf{A}) = \mathbf{B} \cdot \mathbf{D} \times \mathbf{A} + \mathbf{C} \cdot \mathbf{D} \times \mathbf{A} = \mathbf{D} \cdot \mathbf{A} \times \mathbf{B} + \mathbf{D} \cdot \mathbf{A} \times \mathbf{C}$. Letting **D** be **i**, **j** and **k**, we see that this identity holds true for each component of **A**, so $\mathbf{A} \times (\mathbf{B}+\mathbf{C}) = \mathbf{A} \times \mathbf{B} + \mathbf{A} \times \mathbf{C}$.

1.14 Vector Identities

1. Derive the identity

$$(\mathbf{A} \times \mathbf{B}) \times (\mathbf{C} \times \mathbf{D}) = [\mathbf{A},\mathbf{B},\mathbf{D}]\mathbf{C} - [\mathbf{A},\mathbf{B},\mathbf{C}]\mathbf{D} \quad (\textbf{\textit{NOTE:}} \text{ This is not an identity! The correct identity reads } (\mathbf{A} \times \mathbf{B}) \times (\mathbf{C} \times \mathbf{D}) = [\mathbf{A},\mathbf{C},\mathbf{D}]\mathbf{B} - [\mathbf{B},\mathbf{C},\mathbf{D}]\mathbf{A}, \text{ so this is what we will prove.})$$

Solution: Using identity 1.31 $(\mathbf{A} \times \mathbf{B}) \times \mathbf{P} = (\mathbf{A} \cdot \mathbf{P})\mathbf{B} - (\mathbf{B} \cdot \mathbf{P})\mathbf{A}$, and writing $\mathbf{P} = \mathbf{C} \times \mathbf{D}$, we get the result $(\mathbf{A} \times \mathbf{B}) \times (\mathbf{C} \times \mathbf{D}) = [\mathbf{A},\mathbf{C},\mathbf{D}]\mathbf{B} - [\mathbf{B},\mathbf{C},\mathbf{D}]\mathbf{A}$ immediately.

3. Derive the identity
$$\mathbf{A} \times (\mathbf{B} \times \mathbf{C}) + \mathbf{B} \times (\mathbf{C} \times \mathbf{A}) + \mathbf{C} \times (\mathbf{A} \times \mathbf{B}) = 0$$

Solution: Use identity 1.30 $\mathbf{A} \times (\mathbf{B} \times \mathbf{C}) = (\mathbf{A} \cdot \mathbf{C})\mathbf{B} - (\mathbf{A} \cdot \mathbf{B})\mathbf{C}$ three times to get

$$\mathbf{A}\times(\mathbf{B}\times\mathbf{C})+\mathbf{B}\times(\mathbf{C}\times\mathbf{A})+\mathbf{C}\times(\mathbf{A}\times\mathbf{B}) = (\mathbf{A}\cdot\mathbf{C})\mathbf{B}-(\mathbf{A}\cdot\mathbf{B})\mathbf{C}+(\mathbf{B}\cdot\mathbf{A})\mathbf{C}-(\mathbf{B}\cdot\mathbf{C})\mathbf{A}+(\mathbf{C}\cdot\mathbf{B})\mathbf{A}-(\mathbf{C}\cdot\mathbf{A})\mathbf{B}.$$

Because the dot product commutes, the right hand side cancels in pairs, so we have the identity.

5. If the vector ω in figure 1.34 is constant, then the acceleration of a particle with position vector **R** is $\mathbf{a} = \omega \times (\omega \times \mathbf{R})$. Simplify this expression.
Solution: Writing this out by components, we have

$$\omega \times (\omega \times \mathbf{R}) = \omega \times \begin{vmatrix} \mathbf{i} & \mathbf{j} & \mathbf{k} \\ \omega_1 & \omega_2 & \omega_3 \\ x & y & z \end{vmatrix} = \begin{vmatrix} \mathbf{i} & \mathbf{j} & \mathbf{k} \\ \omega_1 & \omega_2 & \omega_3 \\ \omega_2 z - \omega_3 y & \omega_3 x - \omega_1 z & \omega_1 y - \omega_2 x \end{vmatrix} =$$

$$= [-(\omega_2^2+\omega_3^2)x+\omega_1\omega_2 y+\omega_1\omega_3 z]\mathbf{i}+[-(\omega_1^2+\omega_3^2)y+\omega_2\omega_1 x+\omega_2\omega_3 z]\mathbf{j}+[-(\omega_1^2+\omega_2^2)z+\omega_3\omega_1 x+\omega_3\omega_2 y]\mathbf{k}.$$

Adding and subtracting terms to make the expression symmetric, we get
$[-(\omega_1^2 + \omega_2^2 + \omega_3^2)x + \omega_1\omega_1 x + \omega_1\omega_2 y + \omega_1\omega_3 z]\mathbf{i}+$
$[-(\omega_1^2 + \omega_2^2 + \omega_3^2)x + \omega_2\omega_1 y + \omega_2\omega_2 y + \omega_2\omega_3 z]\mathbf{j}+$
$[-(\omega_1^2 + \omega_2^2 + \omega_3^2)x + \omega_3\omega_1 z + \omega_3\omega_2 y + \omega_3\omega_3 z]\mathbf{k}.$
This can be written as $(\omega_1 x + \omega_2 y + \omega_3 z)(\omega_1\mathbf{i} + \omega_2\mathbf{j} + \omega_3\mathbf{k}) - (\omega_1^2 + \omega_2^2 + \omega_3^2)(x\mathbf{i} + y\mathbf{j} + z\mathbf{k})$, or $(\omega \cdot \mathbf{R})\omega - (\omega \cdot \omega)\mathbf{R}$.

7. Simplify $|\mathbf{A} \times \mathbf{B}|^2 + (\mathbf{A} \cdot \mathbf{B})^2 - |\mathbf{A}|^2|\mathbf{B}|^2$.
Solution: Because $|\mathbf{A} \times \mathbf{B}|^2 = |\mathbf{A}|^2|\mathbf{B}|^2 \sin^2 \theta$, and $(\mathbf{A} \cdot \mathbf{B})^2 = |\mathbf{A}|^2|\mathbf{B}|^2 \cos^2 \theta$, we see $|\mathbf{A} \times \mathbf{B}|^2 + |\mathbf{A}|^2|\mathbf{B}|^2 \cos^2 \theta = |\mathbf{A}|^2|\mathbf{B}|^2$, so the expression $|\mathbf{A} \times \mathbf{B}|^2 + (\mathbf{A} \cdot \mathbf{B})^2 - |\mathbf{A}|^2|\mathbf{B}|^2$ sums to zero.

9. Prove, for any vector **A**, that

$$\mathbf{i} \times (\mathbf{i} \times \mathbf{A}) + \mathbf{j} \times (\mathbf{j} \times \mathbf{A}) + \mathbf{k} \times (\mathbf{k} \times \mathbf{A}) = -2\mathbf{A}$$

1.15. TENSOR NOTATION

Solution:
$$\mathbf{i} \times (\mathbf{i} \times \mathbf{A}) + \mathbf{j} \times (\mathbf{j} \times \mathbf{A}) + \mathbf{k} \times (\mathbf{k} \times \mathbf{A}) =$$

$$= \mathbf{i} \times \begin{vmatrix} \mathbf{i} & \mathbf{j} & \mathbf{k} \\ 1 & 0 & 0 \\ A_1 & A_2 & A_3 \end{vmatrix} + \mathbf{j} \times \begin{vmatrix} \mathbf{i} & \mathbf{j} & \mathbf{k} \\ 0 & 1 & 0 \\ A_1 & A_2 & A_3 \end{vmatrix} + \mathbf{k} \times \begin{vmatrix} \mathbf{i} & \mathbf{j} & \mathbf{k} \\ 0 & 0 & 1 \\ A_1 & A_2 & A_3 \end{vmatrix} =$$

$$= \begin{vmatrix} \mathbf{i} & \mathbf{j} & \mathbf{k} \\ 1 & 0 & 0 \\ 0 & -A_3 & A_2 \end{vmatrix} + \begin{vmatrix} \mathbf{i} & \mathbf{j} & \mathbf{k} \\ 0 & 1 & 0 \\ A_3 & 0 & -A_1 \end{vmatrix} + \begin{vmatrix} \mathbf{i} & \mathbf{j} & \mathbf{k} \\ 0 & 0 & 1 \\ -A_2 & A_1 & 0 \end{vmatrix} =$$

$$= (-A_2\mathbf{j} - A_3\mathbf{k}) + (-A_1\mathbf{i} - A_3\mathbf{k}) + (-A_1\mathbf{i} - A_2\mathbf{j}) = -2\mathbf{A}$$

11. Simplify $[\mathbf{A} \times (\mathbf{A} \times \mathbf{B})] \times \mathbf{A} \cdot \mathbf{C}$.
    Solution: Let $\mathbf{A} \times \mathbf{B} = \mathbf{N} = |\mathbf{A}||\mathbf{B}|\sin\theta\,\mathbf{n}$, where θ is the angle between $\mathbf{A}$ and $\mathbf{B}$, and $\mathbf{n} = \frac{\mathbf{N}}{|\mathbf{N}|}$. Then $[\mathbf{A} \times (\mathbf{A} \times \mathbf{B})] \times \mathbf{A} \cdot \mathbf{C} = [\mathbf{A} \times \mathbf{N}] \times \mathbf{A} \cdot \mathbf{C}$. Then write $\mathbf{A} \times \mathbf{N} = \mathbf{M}$, and because $\mathbf{M}$ is perpendicular to both $\mathbf{A}$ and $\mathbf{N}$, $\mathbf{M} = |\mathbf{A}||\mathbf{A}||\mathbf{B}|\sin\theta\,\mathbf{m}$ where $\mathbf{m} = \frac{\mathbf{M}}{|\mathbf{M}|}$. Then we have $\mathbf{M} \times \mathbf{A} \cdot \mathbf{C} = |\mathbf{A}||\mathbf{A}||\mathbf{A}||\mathbf{B}|\sin\theta\,\mathbf{n} \cdot \mathbf{C} = |\mathbf{A}|^2 |\mathbf{A}||\mathbf{B}|\sin\theta\frac{\mathbf{A}\times\mathbf{B}}{|\mathbf{A}\times\mathbf{B}|} \cdot \mathbf{C} = |\mathbf{A}|^2 \mathbf{A} \times \mathbf{B} \cdot \mathbf{C}$. Thus $[\mathbf{A} \times (\mathbf{A} \times \mathbf{B})] \times \mathbf{A} \cdot \mathbf{C} = |\mathbf{A}|^2 [\mathbf{A}, \mathbf{B}, \mathbf{C}]$.

13. Prove the following theorem of Desargues. Given two (nondegenerate) triangles ABC and DEF with the property that the line through AD, the line through BE, and the line through CF have a point in common; moreover, let the lines through AB and DE intersect at P, the lines through BC and EF intersect at Q, and the lines through AC and DF intersect at R. Then P, Q, and R are collinear.
    Solution: Note that the statement of the theorem does not specify that the triangles lie in the same plane: in fact, it is easiest to prove this theorem in three dimensions. Referring to Figure 1.5, if the lines joining the vertices of triangles ABC and DEF meet at a single point O, then one of the triangles, say ABC forms the base of a tetrahedron with the line segments OA, OB and OC, and the triangle DEF lies in this tetrahedron. Assuming the planes M and N containing ABC and DEF are not parallel, these planes meet in a line L. Let the planes containing the three other sides of the tetrahedron be denoted T_1, T_2 and T_3. Each of these planes contains a pair of lines coincident with pairs of sides of the two triangles, and we denote these lines by the lower case version of the notation for the sides of the triangles they contain: T_1 contains ab and de, T_2 contains bc and ef and T_3 contains ca and fd. We now have everything we need to prove the theorem.

    (a) T_1 and M intersect in ab, T_1 and N intersect in de, and T_1, M and N all intersect at the point P in the line L. Thus ab and de intersect at P in L.

    (b) T_2 and M intersect in bc, T_2 and N intersect in ef, and T_2, M and N all intersect at the point Q in the line L. Thus bc and ef intersect at Q in L.

    (c) T_3 and M intersect in ca, T_3 and N intersect in fd, and T_3, M and N all intersect at the point R in the line L. Thus ca and fd intersect at R in L.

    In the case of the triangles being in parallel planes, but not in the same plane, the lines along their sides meet in the so called "line at infinity", and the analysis above is still valid, but the justification is based on ideas not in the book. In the case of co-planar triangles, we can think of this case as the limit of a sequence of constructions in three space that approaches the two dimensional construction in the limit, and again the results above will hold.

15. Write $(\mathbf{u} \times \mathbf{v}) \cdot (\mathbf{u} \times \mathbf{v})$ as a determinant involving only scalar products.
    Solution: Using identity 1.33 $(\mathbf{A} \times \mathbf{B}) \cdot (\mathbf{C} \times \mathbf{D}) = (\mathbf{A} \cdot \mathbf{C})(\mathbf{B} \cdot \mathbf{D}) - (\mathbf{A} \cdot \mathbf{D})(\mathbf{B} \cdot \mathbf{C})$, substitute $\mathbf{u}$ and $\mathbf{v}$ to get $(\mathbf{u} \times \mathbf{v}) \cdot (\mathbf{u} \times \mathbf{v}) = (\mathbf{u} \cdot \mathbf{u})(\mathbf{v} \cdot \mathbf{v}) - (\mathbf{u} \cdot \mathbf{v})(\mathbf{u} \cdot \mathbf{v})$. We see that we can write this as a determinant $\begin{vmatrix} \mathbf{u} \cdot \mathbf{u} & \mathbf{u} \cdot \mathbf{v} \\ \mathbf{u} \cdot \mathbf{v} & \mathbf{v} \cdot \mathbf{v} \end{vmatrix}$.

1.15 Tensor Notation

1. Simplify $(\mathbf{A} \times \mathbf{B}) \cdot \mathbf{C}$.
    Solution: $(\mathbf{A} \times \mathbf{B}) \cdot \mathbf{C} = (\epsilon_{ijk} A_j B_k) C_i = \epsilon_{jki} A_j B_k C_i = [\mathbf{A}, \mathbf{B}, \mathbf{C}]$.

1.15. TENSOR NOTATION

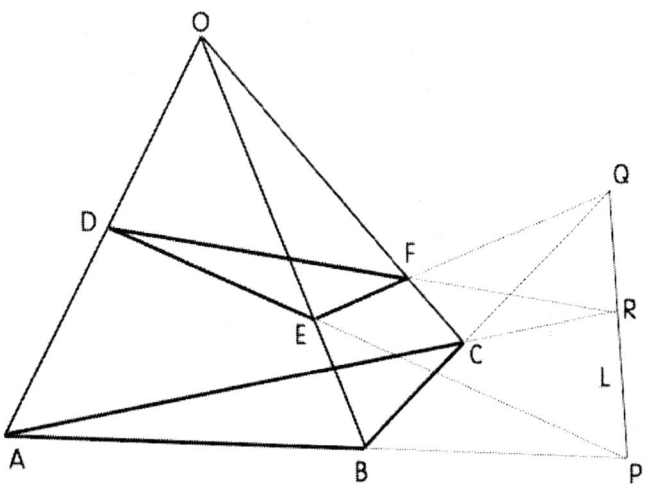

Figure 1.5: The triangles ABC and DEF with the important lines.

3. Simplify $(\mathbf{A} \times \mathbf{B}) \cdot (\mathbf{B} \times \mathbf{C}) \times (\mathbf{C} \times \mathbf{A})$.
   Solution:

$$\begin{aligned}
(\mathbf{A} \times \mathbf{B}) \cdot (\mathbf{B} \times \mathbf{C}) \times (\mathbf{C} \times \mathbf{A}) &= (\mathbf{A} \times \mathbf{B})_i [(\mathbf{B} \times \mathbf{C}) \times (\mathbf{C} \times \mathbf{A})]_i \\
&= (\epsilon_{iuv} A_u B_v) \epsilon_{ijk} (\epsilon_{jpq} B_p C_q)(\epsilon_{krs} C_r A_s) \\
&= \epsilon_{iuv} \epsilon_{ijk} \epsilon_{jpq} \epsilon_{krs} A_u B_v B_p C_q C_r A_s \\
&= (\delta_{uj} \delta_{vk} - \delta_{uk} \delta_{jv}) \epsilon_{jpq} \epsilon_{krs} A_u B_v B_p C_q C_r A_s \\
&= \delta_{uj} \delta_{vk} \epsilon_{jpq} \epsilon_{krs} A_u B_v B_p C_q C_r A_s - \delta_{uk} \delta_{jv} \epsilon_{jpq} \epsilon_{krs} A_u B_v B_p C_q C_r A_s \\
&= \epsilon_{upq} \epsilon_{vrs} A_u B_v B_p C_q C_r A_s - \epsilon_{vpq} \epsilon_{urs} A_u B_v B_p C_q C_r A_s \\
&= [\mathbf{A}, \mathbf{B}, \mathbf{C}][\mathbf{B}, \mathbf{C}, \mathbf{A}] - [\mathbf{B}, \mathbf{B}, \mathbf{C}][\mathbf{A}, \mathbf{C}, \mathbf{A}] \\
&= [\mathbf{A}, \mathbf{B}, \mathbf{C}][\mathbf{A}, \mathbf{B}, \mathbf{C}] - [\mathbf{B}, \mathbf{B}, \mathbf{C}][\mathbf{A}, \mathbf{A}, \mathbf{C}] \\
&= [\mathbf{A}, \mathbf{B}, \mathbf{C}]^2
\end{aligned}$$

1.15. TENSOR NOTATION

Chapter 2

Vector Functions of a Single Variable

2.1 Differentiation

1. Let $\mathbf{F}(t) = \sin t \mathbf{i} + \cos t \mathbf{j} + \mathbf{k}$.

    a) Find $\mathbf{F}'(t)$.
    b) Show that $\mathbf{F}'(t)$ is always parallel to the xy plane.
    c) For what values of t is $\mathbf{F}'(t)$ parallel to the xz plane?
    d) Does $\mathbf{F}(t)$ have constant magnitude?
    e) Does $\mathbf{F}'(t)$ have constant magnitude?
    f) Compute $\mathbf{F}''(t)$.

    Solution:
    a) $\mathbf{F}'(t) = \frac{d\mathbf{F}}{dt} = \cos t \mathbf{i} - \sin t \mathbf{j}$
    b) $\mathbf{F}'(t) \cdot \mathbf{k} = 0$, so there is no z component.
    c) $F'(t) \cdot \mathbf{j} = -\sin t$ which is 0 for $t = \pm n\pi$, and for these values there is no y component.
    d) $|\mathbf{F}(t)| = \sqrt{\sin^2 t + \cos^2 t + 1} = \sqrt{1+1} = \sqrt{2}$, so yes, $\mathbf{F}(t)$ has constant magnitude.
    e) $|\mathbf{F}'(t)| = \sqrt{\cos^2 t + (-\sin t)^2} = \sqrt{1}$, so yes, $\mathbf{F}'(t)$ has constant magnitude.
    f) $\mathbf{F}''(t) = \frac{d\mathbf{F}'(t)}{dt} = -\sin t \mathbf{i} - \cos t \mathbf{j}$

3. Find $f'(t)$ in each of the following cases:

    a) $f(t) = (3t\mathbf{i} + 5t^2\mathbf{j}) \cdot (t\mathbf{i} - \sin t \mathbf{j})$
    b) $f(t) = |2t\mathbf{i} + 2t\mathbf{j} - \mathbf{k}|$
    c) $f(t) = [(\mathbf{i} + \mathbf{j} - 2\mathbf{k}) \times (3t^4\mathbf{i} + t\mathbf{j})] \cdot \mathbf{k}$

    Solution:
    a) $f(t) = (3t\mathbf{i} + 5t^2\mathbf{j}) \cdot (t\mathbf{i} - \sin t \mathbf{j}) = 3t^2 - 5t^2 \sin t$, so $f'(t) = \frac{d}{dt}(3t^2 - 5t^2 \sin t) = 6t - 10t \sin t - 5t^2 \cos t$.
    b) $f(t) = |2t\mathbf{i} + 2t\mathbf{j} - \mathbf{k}| = \sqrt{8t^2 + 1}$ so $f'(t) = \frac{d}{dt}\sqrt{8t^2+1}$ or $f'(t) = \frac{8t}{\sqrt{8t^2+1}}$.
    c) $f(t) = [(\mathbf{i}+\mathbf{j}-2\mathbf{k}) \times (3t^4\mathbf{i}+t\mathbf{j})] \cdot \mathbf{k} = [2t\mathbf{i} - 6t^4\mathbf{j} + (t-3t^4)\mathbf{k}] \cdot \mathbf{k} = t - 3t^4$ so $f'(t) = \frac{d}{dt}(t - 3t^4)$
    $f'(t) = 1 - 12t^3$.

2.2. SPACE CURVES VELOCITIES AND TANGENTS

5. Given the three vectors $\mathbf{A} = 3\mathbf{i} + 2\mathbf{j} + 6\mathbf{k}$ $\mathbf{B} = 3\mathbf{i} + 4\mathbf{k}$, and $\mathbf{C} = 2\mathbf{i} - 2\mathbf{j} + \mathbf{k}$, evaluate
   a) $|\mathbf{A}|$
   b) $\mathbf{A} \cdot \mathbf{B}$
   c) $\mathbf{B} \times \mathbf{C}$
   d) $\mathbf{B} \cdot \mathbf{B} \times \mathbf{C}$
   e) $[\mathbf{A}, \mathbf{B}, \mathbf{C}]$
   f) $\mathbf{A}/|\mathbf{B}|$
   g) $\mathbf{A} \times (\mathbf{B} \times \mathbf{C})$
   h) $\frac{d}{dt}(\mathbf{A} + \mathbf{B}t)$
   i) $\frac{d}{dt}(\mathbf{B} \times t\mathbf{C})$

   Solution:
   a) $|\mathbf{A}| = \sqrt{3^2 + 2^2 + 6^2} = \sqrt{49} = 7$
   b) $\mathbf{A} \cdot \mathbf{B} = 3 \cdot 3 + 6 \cdot 4 = 9 + 24 = 33$
   c) $\mathbf{B} \times \mathbf{C} = 8\mathbf{i} + 5\mathbf{j} - 6\mathbf{k}$
   d) $\mathbf{B} \cdot \mathbf{B} \times \mathbf{C} = (3\mathbf{i} + 4\mathbf{k}) \cdot (8\mathbf{i} + 5\mathbf{j} - 6\mathbf{k}) = 24 - 24 = 0$
   e) $[\mathbf{A}, \mathbf{B}, \mathbf{C}] = \mathbf{A} \cdot (\mathbf{B} \times \mathbf{C}) = (3\mathbf{i} + 2\mathbf{j} + 6\mathbf{k}) \cdot (8\mathbf{i} + 5\mathbf{j} - 6\mathbf{k}) = -2$
   f) $\mathbf{A}/|\mathbf{B}| = (3\mathbf{i} + 2\mathbf{j} + 6\mathbf{k})/\sqrt{25} = \frac{3}{5}\mathbf{i} + \frac{2}{5}\mathbf{j} + \frac{6}{5}\mathbf{k}$
   g) $\mathbf{A} \times (\mathbf{B} \times \mathbf{C}) = (3\mathbf{i} + 2\mathbf{j} + 6\mathbf{k}) \times (8\mathbf{i} + 5\mathbf{j} - 6\mathbf{k}) = -42\mathbf{i} + 66\mathbf{j} - 1\mathbf{k}$
   h) $\frac{d}{dt}(\mathbf{A} + \mathbf{B}t) = \frac{d}{dt}((3 + 3t)\mathbf{i} + 2\mathbf{j} + (6 + 4t)\mathbf{k}) = 3\mathbf{i} + 4\mathbf{k} = \mathbf{B}$
   i) $\frac{d}{dt}(\mathbf{B} \times t\mathbf{C}) = \frac{d}{dt}(8t\mathbf{i} + 5t\mathbf{j} - 6t\mathbf{k}) = 8\mathbf{i} + 5\mathbf{j} - 6\mathbf{k}$

2.2 Space Curves Velocities and Tangents

1. Find the unit vector tangent to the oriented closed curve
   $x = a\cos t, y = b\sin t, z = 0$ at $t = \frac{3}{2}\pi$.
   Solution:

$$\mathbf{T} = \frac{(dx/dt)\mathbf{i} + (dy/dt)\mathbf{j} + (dz/dt)\mathbf{k}}{\sqrt{(dx/dt)^2 + (dy/dt)^2 + (dz/dt)^2}}$$

$$\mathbf{T}\left(\frac{3}{2}\pi\right) = \frac{(-a\sin(\frac{3}{2}\pi))\mathbf{i} + (b\cos(\frac{3}{2}\pi))\mathbf{j}}{\sqrt{(a\sin(\frac{3}{2}\pi))^2 + (b\cos(\frac{3}{2}\pi))^2}} = \frac{a\mathbf{i}}{\sqrt{(a)^2}} = \mathbf{i}$$

3. Observe that
$$x = \frac{t}{2\pi}, \quad y = \sin t, \quad z = \cos t$$

is a parametrization of the helix in example 2.14.
Compute the arc length between the same two endpoints using formula (2.24).
What is the unit tangent vector at $(0, 0, 1)$?

Solution:

$$\int_a^b \left|\frac{d\mathbf{R}}{dt}\right| dt = \text{Length of Curve}$$

$$= \int_0^1 \left|\frac{d}{dt}\left((t/2\pi)\mathbf{i} + \sin t\mathbf{j} + \cos t\mathbf{k}\right)\right| dt$$

$$= \int_0^1 \sqrt{\left(\frac{1}{2\pi}\right)^2 + (\cos t)^2 + (-\sin t)^2}\, dt$$

$$= \int_0^1 \sqrt{\left(\frac{1}{2\pi}\right)^2 + 1}\, dt$$

$$= \sqrt{\left(\frac{1}{2\pi}\right)^2 + 1}$$

The tangent vector is $\frac{1}{2\pi}\mathbf{i} + \cos t\mathbf{j} - \sin t\mathbf{k}$, the unit tangent vector is $\frac{1}{\sqrt{1+\left(\frac{1}{2\pi}\right)^2}}(\frac{1}{2\pi}\mathbf{i} + \cos t\mathbf{j} - \sin t\mathbf{k})$ and the helix passes through $(0,0,1)$ when $t = 0$, so the unit tangent at that point is

$$\frac{\frac{1}{2\pi}\mathbf{i} + \mathbf{j}}{\sqrt{1 + \left(\frac{1}{2\pi}\right)^2}}.$$

5. a) Determine the arc length of the curve

$$x = e^t \cos t,\ y = e^t \sin t,\ z = 0$$

between $t = 0$ and $t = 1$.

  b) Reparameterize the curve in terms of arc length.

  c) This curve is a spiral. Sketch it to see why.

Solution:

a) The arc length is $s = \int_0^t |(e^t \cos t - e^t \sin t)\mathbf{i} + (e^t \sin t + e^t \cos t)\mathbf{j}|\, dt = \int_0^t \sqrt{2}e^t dt = \sqrt{2}(e^t - 1)$, and for an upper limit of $t = 1$, this is $\sqrt{2}(e - 1)$.

b) To reparameterize the curve in terms of arc length, we must find s in terms of t and substitute into the formula for the curve. Because $s = \sqrt{2}(e^t - 1)$, $t = \ln\left(\frac{s}{\sqrt{2}} + 1\right)$, and the curve is $\left(\frac{s}{\sqrt{2}} + 1\right)\left(\cos\ln\left(\frac{s}{\sqrt{2}} + 1\right)\mathbf{i} + \sin\ln\left(\frac{s}{\sqrt{2}} + 1\right)\mathbf{j}\right)$.

c) It is easy to see that the formula for the original curve describes circular motion with an exponentially growing radius, or a spiral. The reparameterized curve is the same shape, but as s changes, we move more slowly along the curve.

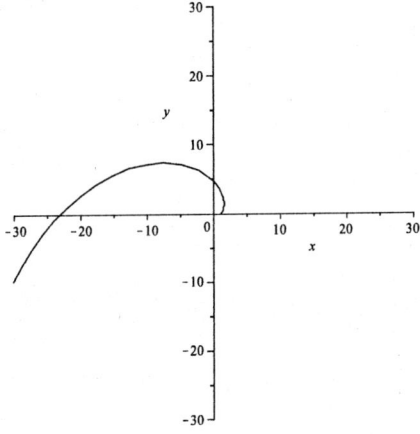

Figure 2.1: The logarithmic spiral of problem 5.

2.3. ACCELERATION AND CURVATURE

7. By using the identities concerning hyperbolic functions, eliminate the parameter t from the equations
$$x = \cosh t, y = \sinh t, z = 0$$
Solution: Because $\cosh t = \frac{1}{2}(e^t + e^{-t})$ and $\sinh t = \frac{1}{2}(e^t - e^{-t})$, we can form the sum and differences of these: $x + y = \cosh t + \sinh t = \frac{1}{2}(e^t + e^{-t}) + \frac{1}{2}(e^t - e^{-t}) = e^t$ and $x - y = \cosh t - \sinh t = \frac{1}{2}(e^t + e^{-t}) - \frac{1}{2}(e^t - e^{-t}) = e^{-t}$. The product $(x+y)(x-y) = e^t e^{-t} = 1$, so we have $x^2 - y^2 = 1$ and $z = 0$.

9. As t varies from -1 to 1, the point (x, y, z) where $x = t$, $y = |t|$, $z = 0$ traces a regular curve. At what point on this curve is there no tangent?
Solution: The tangent to the curve is not defined at $t = 0$, because $(0, 0, 0)$ is a corner.

11. Is the helix of example 2.14 right- or left-handed?
Solution: In a right handed coordinate system, looking at the origin from the first quadrant $\mathbf{e}_1$, $\mathbf{e}_2$ and $\mathbf{e}_3$ appear in this order counterclockwise. In the coordinate system given, where $\mathbf{e}_1$, $\mathbf{e}_2$, $\mathbf{e}_3$ are along the z, y, x axes, respectively, they appear in this order counterclockwise. Because an increase along the x axis is a positive change, by the conditions mentioned in the text around equation (2.15), it is a left handed helix.

13. Suppose that $P_1 P_2$ is a smooth arc in the xy plane.
Is it necessarily true that dy/dx exists at every point on this arc?
Solution: No, because the tangent vector can be vertical, in which case it is not defined.

15. Show that the graph of any continuously differentiable function $y = f(x)$ is a smooth curve. [Hint: Check the parametrization $x = t$, $y = f(t)$, $z = 0$.]
Solution: The three conditions guaranteeing a function being smooth are listed on page 78 of the text. Because we are told the function is continuously differentiable, it is also continuous. Since it is continuously differentiable it cannot turn around so that it must be single valued over the interval for which it is continuously differentiable. Finally, because our function is continuously differentiable, $\frac{d\mathbf{R}}{dt}$ can never equal zero, so all three conditions have been met.

2.3 Acceleration and Curvature

In problems 1-4, the coordinates of a moving particle are given as a function of time t. Find (a) the speed, (b) the tangential and normal components of acceleration, (c) the unit tangent vector $\mathbf{T}$, and (d) the curvature of the curve, as functions of time.

1. $x = e^t \cos t, y = e^t \sin t, z = 0$
Solution:

- The speed is the magnitude of the velocity $\frac{d\mathbf{R}}{dt}$ or $\left|\frac{d\mathbf{R}}{dt}\right| = \sqrt{(\frac{dx}{dt})^2 + (\frac{dy}{dt})^2 + (\frac{dz}{dt})^2}$.
- The tangential component of acceleration a_t is the derivative of the speed $\frac{d}{dt}\left|\frac{d\mathbf{R}}{dt}\right| = \frac{d^2 s}{dt^2}$. One way to find the normal component is by solving $a = (a_n^2 + a_t^2)^{1/2}$ for a_n.
- The unit tangent vector $\mathbf{T}$ is the unit velocity vector $\frac{\mathbf{v}}{|\mathbf{v}|}$
- The curvature can be found by several methods, one of which is to compute $k = \frac{|\mathbf{R}' \times \mathbf{R}''|}{|\mathbf{R}'|^3}$.

(a) The velocity is $\mathbf{v} = e^t(\cos t - \sin t)\mathbf{i} + e^t(\sin t + \cos t)\mathbf{j}$ and the magnitude of the velocity (the speed) is $\sqrt{2} e^t$.

(b) The tangential component of acceleration is $\frac{d}{dt}\sqrt{2}e^t = \sqrt{2}e^t$. The acceleration $\mathbf{a}$ is $2e^t(-\sin t \mathbf{i} + \cos t \mathbf{j})$, the magnitude a is $2e^t$, so the normal component is $a_n = (a^2 - a_t^2)^{1/2} = (4e^{2t} - 2e^{2t})^{1/2} = \sqrt{2}e^t$.

(c) We already found the velocity $\mathbf{v} = e^t(\cos t - \sin t)\mathbf{i} + e^t(\sin t + \cos t)\mathbf{j}$ so divide by its magnitude $\sqrt{2}e^t$ to get $\frac{\sqrt{2}}{2}((\cos t - \sin t)\mathbf{i} + (\sin t + \cos t)\mathbf{j})$.

2.3. ACCELERATION AND CURVATURE

(d) We have all we need to compute the curvature: $\mathbf{R}'$, $\mathbf{R}''$ and $|\mathbf{R}'|$, so putting these together
$k = \frac{|\mathbf{R}' \times \mathbf{R}''|}{|\mathbf{R}'|^3} = \frac{2e^{2t}}{2\sqrt{2}e^{3t}} = \frac{\sqrt{2}}{2}e^{-t}$.

3. $x = e^t \cos t$, $y = e^t \sin t$, $z = e^t$
   Solution:

   (a) Now the velocity is $\mathbf{v} = e^t((\cos t - \sin t)\mathbf{i} + (\sin t + \cos t)\mathbf{j} + \mathbf{k})$ and the magnitude of the velocity (the speed) is $\sqrt{3}e^t$.

   (b) The tangential component of acceleration is $\frac{d}{dt}\sqrt{3}e^t = \sqrt{3}e^t$. The acceleration $\mathbf{a}$ is $2e^t(-\sin t\,\mathbf{i} + \cos t\,\mathbf{j}) + e^t\mathbf{k}$, the magnitude a is $\sqrt{5}e^t$, so the normal component is $a_n = (a^2 - a_t^2)^{1/2} = (5e^{2t} - 3e^{2t})^{1/2} = \sqrt{2}e^t$.

   (c) Take the velocity $\mathbf{v} = e^t((\cos t - \sin t)\mathbf{i} + (\sin t + \cos t)\mathbf{j} + \mathbf{k})$ and divide by its magnitude $\sqrt{3}e^t$ to get $\frac{\sqrt{3}}{3}((\cos t - \sin t)\mathbf{i} + (\sin t + \cos t)\mathbf{j} + \mathbf{k})$.

   (d) Again, we have $\mathbf{R}'$, $\mathbf{R}''$ and $|\mathbf{R}'|$, so compute $k = \frac{|\mathbf{R}' \times \mathbf{R}''|}{|\mathbf{R}'|^3} = \frac{\sqrt{6}e^{2t}}{3\sqrt{3}e^{3t}} = \frac{\sqrt{2}}{3}e^{-t}$.

5. The position vector of a moving particle is
$$\mathbf{R} = \cos t(\mathbf{i} - \mathbf{j}) + \sin t(\mathbf{i} + \mathbf{j}) + \tfrac{1}{2}t\mathbf{k}$$

   (a) Determine the velocity and the speed of the particle.
   (b) Determine the acceleration of the particle.
   (c) Find a unit tangent to the path of the particle, in the direction of motion.
   (d) Show that the curve traversed by the particle has constant curvature k, and find its value.
   (e) Show that the curve is a helix.

   Solution:

   (a) The velocity is $(\cos t - \sin t)\mathbf{i} + (\cos t + \sin t)\mathbf{j} + 1/2\,\mathbf{k}$ and the speed $3/2$, is the magnitude of the velocity.

   (b) The acceleration is $(-\cos t - \sin t)\mathbf{i} + (\cos t - \sin t)\mathbf{j} + 0\,\mathbf{k}$.

   (c) The unit tangent to the path is $\frac{\mathbf{v}}{|\mathbf{v}|} = \mathbf{T} = \frac{2}{3}[(\cos t - \sin t)\mathbf{i} + (\cos t + \sin t)\mathbf{j} + 1/2\,\mathbf{k}]$.

   (d) The curvature $k = \frac{|\mathbf{R}' \times \mathbf{R}''|}{|\mathbf{R}'|^3} = 4/9\sqrt{2}$, which is constant.

   (e) To show that the curve is a helix, we must find an equivalence between $(\sin t + \cos t)\mathbf{i} + (\sin t - \cos t)\mathbf{j} + 1/2t\,\mathbf{k}$ and the general form of the helix $a\cos(\omega t + \gamma)\mathbf{i} + b\sin(\omega t + \gamma)\mathbf{j} + ct\,\mathbf{k}$ for some values of a, b, c, ω and γ, the phase offset. If we write the amplitudes of the each of the vector components in terms of sines and cosines, we may be able to massage the mystery vector function into a form where we can use a sum of angle identity.
   Label the coefficients of the sine and cosine functions in $(\cos t + \sin t)\mathbf{i} + (-\cos + \sin t)\mathbf{j} + 1/2t\,\mathbf{k}$ by A, B so that $A = 1, B = 1$ in the first component and $A = -1, B = 1$ in the second. Then we can write $\mathbf{R}(t) = \sqrt{2}(\frac{1}{\sqrt{2}}\cos t + \frac{1}{\sqrt{2}}\sin t)\mathbf{i} + \sqrt{2}(-\frac{1}{\sqrt{2}}\cos + \frac{1}{\sqrt{2}}\sin t)\mathbf{j} + 1/2t\,\mathbf{k}$.
   From figure 2.2 we can see that $\cos\alpha = \frac{1}{\sqrt{2}}$ and $\sin\alpha = \frac{1}{\sqrt{2}}$, so we write $\mathbf{R}(t) = \sqrt{2}(\cos\alpha\cos t + \sin\alpha\sin t)\mathbf{i} + \sqrt{2}(\cos\alpha\sin t - \sin\alpha\cos t)\mathbf{j} + 1/2t\,\mathbf{k} = \sqrt{2}(\cos(t - \alpha))\mathbf{i} + \sqrt{2}(\sin t - \alpha)\mathbf{j} + 1/2t\,\mathbf{k}$, which is a helix.
   This is a good trick to know. *Any* linear combination of sines and cosines with the same frequency always form a sine or cosine with a different amplitude and some phase offset, and the constants can be computed just as we did here.

7. Let C be the curve given by the equation
$$\mathbf{R}(t) = \cos t\,\mathbf{i} + \sin t\,\mathbf{j} + \log\sec t\,\mathbf{k}$$

   Find

   (a) the element of arc length, ds, along C, in terms of t.

2.3. ACCELERATION AND CURVATURE

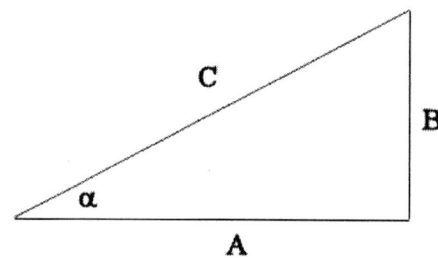

Figure 2.2: Triangle for determining amplitude of cosine and sine functions in equivalent helix.

(b) the unit tangent $\mathbf{T}$.
(c) the unit normal $\mathbf{N}$.
(d) the curvature k.

Solution: Recall that

- Arc length ds is given by $\left|\frac{d\mathbf{R}}{dt}\right| dt$.
- The unit tangent $\mathbf{T}$ is found by computing $\frac{\mathbf{v}}{|\mathbf{v}|}$.
- The unit normal $\mathbf{N}$ is $\frac{d\mathbf{T}}{dt} / \left|\frac{d\mathbf{T}}{dt}\right|$.
- The curvature k can also be computed as $\left|\frac{d\mathbf{T}}{dt} / \frac{ds}{dt}\right|$ or $\left|\frac{d\mathbf{T}}{dt} / \left|\frac{d\mathbf{R}}{dt}\right|\right|$.

So, we find

(a) $ds = \left|\frac{d\mathbf{R}}{dt}\right| dt = |-\sin t\,\mathbf{i} + \cos t\,\mathbf{j} + \tan t\,\mathbf{k}|\,dt = \sqrt{1 + \tan^2 t}\,dt = \sec t\,dt$.

(b) $\mathbf{T} = \frac{d\mathbf{R}}{dt} / \left|\frac{d\mathbf{R}}{dt}\right| = \frac{1}{\sqrt{1+\tan^2 t}}(-\sin t\,\mathbf{i} + \cos t\,\mathbf{j} + \tan t\,\mathbf{k}) = \frac{1}{\sec t}(-\sin t\,\mathbf{i} + \cos t\,\mathbf{j} + \tan t\,\mathbf{k}) = -\sin t \cos t\,\mathbf{i} + \cos^2 t\,\mathbf{j} + \sin t\,\mathbf{k}$.

(c) $\mathbf{N} = \frac{1}{\sqrt{(\sin^2 t - \cos^2 t)^2 + (-2\cos t\sin t)^2 + \cos^2 t}}((\sin^2 t - \cos^2 t)\,\mathbf{i} - 2\cos t \sin t\,\mathbf{j} + \cos t\,\mathbf{k})$
$= \frac{1}{\sqrt{1+\cos^2 t}}((\sin^2 t - \cos^2 t)\,\mathbf{i} - 2\cos t \sin t\,\mathbf{j} + \cos t\,\mathbf{k})$.

(d) Curvature $k = \left|\frac{d\mathbf{T}}{dt} / \sec t\right| = \left|((\sin^2 t - \cos^2 t)\,\mathbf{i} - 2\cos t \sin t\,\mathbf{j} + \cos t\,\mathbf{k})\cos t\right| = \sqrt{\cos^2(\sin^2 t - \cos^2 t)^2 + (-2\cos^2 t \sin t) + \cos^2 t} = \cos t\sqrt{1 + \cos^2 t}$.

9. A particle moves so that its position $\mathbf{R}$ at time t is given by

$$\mathbf{R}(t) = \log(t^2 + 1)\,\mathbf{i} + (t - 2\arctan t)\,\mathbf{j} + 2\sqrt{2}t\,\mathbf{k}$$

(a) Show that this particle moves with constant speed $v = 3$.
(b) Find the curvature of the path of this particle.

Solution:

(a) The velocity is $\frac{d\mathbf{R}}{dt} = \frac{d}{dt}\log(t^2+1)\,\mathbf{i} + \frac{d}{dt}(t - 2\arctan t)\,\mathbf{j} + \frac{d}{dt}2\sqrt{2}t\,\mathbf{k} = (\frac{1}{1+t^2})2t\,\mathbf{i} + (1 - 2\frac{d}{dt}\arctan t)\,\mathbf{j} + 2\sqrt{2}\,\mathbf{k} = \frac{2t}{t^2+1}\,\mathbf{i} + (1 - \frac{2}{t^2+1})\,\mathbf{j} + 2\sqrt{2}\,\mathbf{k}$. The magnitude is

$\sqrt{(\frac{2t}{t^2+1})^2 + (1 - \frac{2}{t^2+1})^2 + (2\sqrt{2})^2} = \sqrt{\frac{4t^2}{(t^2+1)^2} + 1 - \frac{4}{t^2+1} + \frac{4}{(t^2+1)^2} + 8} = \sqrt{\frac{4t^2 - 4(1+t^2) + 4}{(t^2+1)^2} + 9} =$
$\sqrt{\frac{4(1+t^2) - 4(1+t^2)}{(t^2+1)^2} + 9} = 3$, therefore the speed is constant.

(b) To find the curvature of the path we need the acceleration also, which is $-\frac{2(t^2-1)}{(t^2+1)^2}\,\mathbf{i} + \frac{4t}{(t^2+1)^2}\,\mathbf{j}$, and using the formula $k = \frac{|\mathbf{R}' \times \mathbf{R}''|}{|\mathbf{R}'|^3}$ we find, noting that $\left|\frac{d\mathbf{R}}{dt}\right| = 3$ so $\left|\frac{d\mathbf{R}}{dt}\right|^3 = 27$,

$$k = \frac{\left|\frac{-8\sqrt{2}t}{(1+t^2)^2}\,\mathbf{i} - \frac{4\sqrt{2}(t^2-1)}{(1+t^2)^2}\,\mathbf{j} + \frac{2}{1+t^2}\right|}{\left|\frac{d\mathbf{R}}{dt}\right|^3} = \frac{1}{27}\sqrt{\frac{128t^2}{(1+t^2)^4} + \frac{32(t^2-1)^2}{(1+t^2)^4} + \frac{4}{(1+t^2)^2}} =$$

$$= \frac{1}{27}\sqrt{\frac{128t^2 + 32(t^2-1)^2 + 4(1+t^2)^2}{(1+t^2)^4}} = \frac{1}{27}\sqrt{\frac{128t^2 + 32t^4 - 64t^2 + 32 + 4t^4 + 8t^2 + 4}{(1+t^2)^4}} =$$

$$= \frac{1}{27}\sqrt{\frac{36t^4 + 72t^2 + 32}{(1+t^2)^4}} = \frac{2}{9(t^2+1)}.$$

11. Graph the planar curve $y = \sin x$. Without writing equations, demonstrate on the graph how the normal $\mathbf{N}$ jumps discontinuously each time the curve crosses the axis. What is the curvature at these points?
    Solution: See figure 2.3 and imagine the osculating circle on the left rolling down the sine curve, with its radius growing to infinity as its tangent to the curve approaches the axis. You can see that it needs to hop to the other side of the curve at the point where the curve crosses the axis.

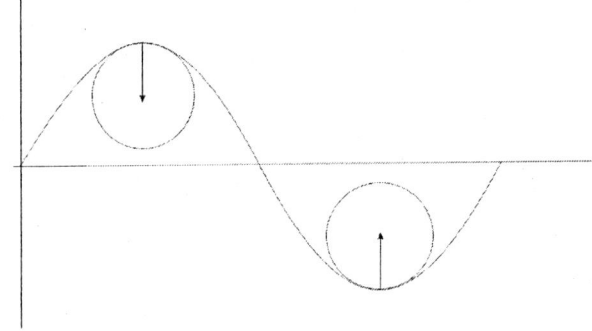

Figure 2.3: The osculating circles must jump from one side of the curve to the other somewhere between the local max and min of the sine curve. We can see that the radius of curvature become infinite as the point of contact of the circle approaches the x axis.

13. Find the curvature and torsion for the helix
$$x = t, \quad y = \sin t, \quad z = \cos t$$

Solution: The curvature can be most easily calculated by the formula $k = \frac{|\mathbf{R}' \times \mathbf{R}''|}{|\mathbf{R}'|^3}$ which gives $1/2$. The torsion is $\tau = -\mathbf{N} \cdot \frac{d\mathbf{B}}{ds}$. To actually calculate this, we need to compute $\mathbf{B} = \mathbf{T} \times \mathbf{N}$ and rewrite $\tau = -\mathbf{N} \cdot \frac{d\mathbf{B}}{ds} = -\mathbf{N} \cdot \frac{d\mathbf{B}}{dt}\frac{dt}{ds} = -\mathbf{N} \cdot \frac{d\mathbf{B}}{dt}/|\frac{d\mathbf{R}}{dt}|$. We find $\mathbf{T} = \frac{\sqrt{2}}{2}(\mathbf{i} + \cos t\,\mathbf{j} - \sin t\,\mathbf{k})$, $\mathbf{N} = -\frac{\sqrt{2}}{2}(\sin t\,\mathbf{j} + \cos t\,\mathbf{k})$, $\mathbf{B} = \mathbf{T} \times \mathbf{N} = \frac{1}{2}(-\mathbf{i} + \cos t\,\mathbf{j} - \sin t\,\mathbf{k})$, $\frac{d\mathbf{B}}{dt} = -\frac{1}{2}(\sin t\,\mathbf{j} + \cos t\,\mathbf{k})$ and $|\frac{d\mathbf{R}}{dt}| = \sqrt{2}$. Then $\tau = -\mathbf{N} \cdot \frac{d\mathbf{B}}{dt}/|\frac{d\mathbf{R}}{dt}| = -1/2$.

15. By inspection, write down the values of each of the following:

    (a) $\frac{d\mathbf{R}}{ds} \cdot \mathbf{T}$
    (b) $\frac{d}{ds}(\mathbf{T} \cdot \mathbf{T})$
    (c) $\frac{d^2\mathbf{R}}{dt^2} \cdot \mathbf{T}$
    (d) $\mathbf{T} \cdot \mathbf{N}$
    (e) $\frac{d\mathbf{R}}{dt} \cdot \mathbf{T}$
    (f) $\frac{d\mathbf{N}}{ds} \cdot \mathbf{B}$
    (g) $[\mathbf{T}, \mathbf{N}, \mathbf{B}]$
    (h) $|\frac{d^2\mathbf{R}}{ds^2}|$
    (i) $|\frac{d\mathbf{B}}{ds}|$

    Solution: In each case, the idea is to use what you know about the properties of the components of the expressions.

2.3. ACCELERATION AND CURVATURE

(a) $\frac{d\mathbf{R}}{ds} \cdot \mathbf{T} = 1$ because these are identical unit vectors.

(b) $\frac{d}{ds}(\mathbf{T} \cdot \mathbf{T}) = 0$, which can be seen in a couple different ways. Because each component of the derivative (using the product rule) is identical and equal to the scalar product $k\mathbf{N} \cdot \mathbf{T}$ or $\mathbf{T} \cdot k\mathbf{N}$ of two orthogonal vectors. Also, the dot product of two identical unit vectors is constant.

(c) $\frac{d^2\mathbf{R}}{dt^2} \cdot \mathbf{T} = \frac{d^2s}{dt^2}$ because $\frac{d^2\mathbf{R}}{dt^2}$ is the acceleration which has components $k(\frac{ds}{dt})^2$ and $\frac{d^2s}{dt^2}$ in the $\mathbf{N}$ and $\mathbf{T}$ directions respectively.

(d) $\mathbf{T} \cdot \mathbf{N} = 0$ because these are orthogonal.

(e) $\frac{d\mathbf{R}}{dt} \cdot \mathbf{T} = \frac{ds}{dt}$ because $\frac{d\mathbf{R}}{dt} = \frac{d\mathbf{R}}{ds}\frac{ds}{dt} = \mathbf{T}\frac{ds}{dt}$.

(f) $\frac{d\mathbf{N}}{ds} \cdot \mathbf{B} = \tau$ because $\frac{d\mathbf{N}}{ds}$ has components k and τ in the $\mathbf{N}$ and $\mathbf{B}$ directions, respectively.

(g) $[\mathbf{T}, \mathbf{N}, \mathbf{B}] = 1$ because $\mathbf{B} = \mathbf{T} \times \mathbf{N}$ by definition.

(h) $|\frac{d^2\mathbf{R}}{ds^2}| = k$ because $\frac{d\mathbf{R}}{ds} = \mathbf{T}$ and $|\frac{d\mathbf{T}}{ds}| = k$.

(i) $\frac{d\mathbf{B}}{ds} = -\tau\mathbf{N}$ by definition.

17. Find the unit tangent $\mathbf{T}$, the principal normal $\mathbf{N}$, the binormal $\mathbf{B}$, the curvature, and the torsion for
$$x = \cos^3 t, \quad y = \sin^3 t, \quad z = 2\sin^2 t, \quad (0 < t \leq \pi/2)$$

Solution: We compute $\mathbf{T} = \frac{d\mathbf{R}}{dt} = -\frac{3}{5}\cos t\,\mathbf{i} + \frac{3}{5}\sin t\,\mathbf{j} + \frac{4}{5}\mathbf{k}$ and $\mathbf{N} = \frac{d\mathbf{T}}{dt} / |\frac{d\mathbf{T}}{dt}| = \frac{3}{5}\sin t\,\mathbf{i} + \frac{3}{5}\cos t\,\mathbf{j}$, and then the

binormal is just $\mathbf{B} = \mathbf{T} \times \mathbf{N} = -\frac{12}{25}\cos t\,\mathbf{i} + \frac{12}{25}\sin t\,\mathbf{j} - \frac{9}{25}\mathbf{k}$. The curvature we calculate from $k = \frac{|\mathbf{R}' \times \mathbf{R}''|}{|\mathbf{R}'|^3} = \frac{3}{25}\frac{1}{\cos t \sin t}$. The torsion τ is given as $\tau = -\mathbf{N} \cdot \frac{d\mathbf{B}}{ds} = -\mathbf{N} \cdot \frac{d\mathbf{B}}{dt}\frac{dt}{ds} = -\mathbf{N} \cdot \frac{d\mathbf{B}}{dt} / |\frac{d\mathbf{R}}{dt}| = -\frac{36}{625}\frac{1}{\sin t \cos t}$.

19. Express the curvature and torsion of a helix $\mathbf{R}(t) = \mathbf{R}_0 + \rho\cos t\,\mathbf{e}_1 + \rho\sin t\,\mathbf{e}_2 + at\,\mathbf{e}_3$ in terms of its radius and pitch.

Solution: The radius is ρ and the pitch is $2a\pi$. Writing the general helix as $\mathbf{R}(t) = (r_1 + \rho\cos t)\mathbf{e}_1 + (r_2 + \rho\sin t)\mathbf{e}_2 + (r_3 + at)\mathbf{e}_3$, we need to compute $\mathbf{R}'$ and $\mathbf{R}''$ to get the curvature, and we need $\mathbf{T}$ and $\mathbf{N}$ to get $\mathbf{B}$ and thus the torsion τ. We find $\mathbf{R}' = -\rho\sin t\,\mathbf{e}_1 + \rho\cos t\,\mathbf{e}_2 + a\,\mathbf{e}_3$, $\mathbf{R}'' = -\rho\cos t\,\mathbf{e}_1 - \rho\sin t\,\mathbf{e}_2$, so $k = \frac{|\mathbf{R}' \times \mathbf{R}''|}{|\mathbf{R}'|^3} = \frac{\rho}{a^2+\rho^2}$. We compute $\mathbf{T} = \frac{1}{\sqrt{a^2+\rho^2}}(-\rho\sin t\,\mathbf{e}_1 + \rho\cos t\,\mathbf{e}_2 + a\,\mathbf{e}_3)$ and $\mathbf{N} = -\cos t\,\mathbf{e}_1 - \sin t\,\mathbf{e}_2$ so $\mathbf{B} = \mathbf{T} \times \mathbf{N} = \frac{1}{\sqrt{a^2+\rho^2}}(a\sin t\,\mathbf{e}_1 - a\cos t\,\mathbf{e}_2 + \rho\,\mathbf{e}_3)$ and the torsion is $\tau = -\mathbf{N} \cdot \frac{d\mathbf{B}}{dt} / |\frac{d\mathbf{R}}{dt}| = \frac{a}{a^2+\rho^2}$.

21. If the curve $\mathbf{R}(t)$ lies on a sphere $|\mathbf{R}(t)| = $ constant, prove that
$$\mathbf{R} = -\rho\mathbf{N} - \frac{1}{\tau}\frac{d\rho}{ds}\mathbf{B}$$

(*Hint*: Keep differentiating $\mathbf{R} \cdot \mathbf{R} = $ constant, using the Frenet formulas.)

Solution: Because $\mathbf{R} \cdot \mathbf{R} = C$, we have $\frac{d\mathbf{R}}{ds} \cdot \mathbf{R} + \mathbf{R} \cdot \frac{d\mathbf{R}}{ds} = 2\frac{d\mathbf{R}}{ds} \cdot \mathbf{R} = 0$. Dividing by 2, substituting $\mathbf{T} = \frac{d\mathbf{R}}{ds}$ and differentiating again, we get $\frac{d\mathbf{T}}{ds} \cdot \mathbf{R} + \mathbf{T} \cdot \frac{d\mathbf{R}}{ds} = k\mathbf{N} \cdot \mathbf{R} + \mathbf{T} \cdot \mathbf{T} = 0$, which means $\mathbf{N} \cdot \mathbf{R} + \rho = 0$, where we have used the fact that $\mathbf{T} \cdot \mathbf{T} = 1$ and $1/k = \rho$. Differentiating again, we get $\frac{d\mathbf{N}}{ds} \cdot \mathbf{R} + \mathbf{N} \cdot \frac{d\mathbf{R}}{ds} = \frac{d\mathbf{N}}{ds} \cdot \mathbf{R} + \mathbf{N} \cdot \mathbf{T} = \frac{d\mathbf{N}}{ds} \cdot \mathbf{R} = -\frac{d\rho}{ds}$. Substitute $\frac{d\mathbf{N}}{ds} = \tau\mathbf{B} - k\mathbf{T}$ to get $(\tau\mathbf{B} - k\mathbf{T}) \cdot \mathbf{R} = \tau\mathbf{B} \cdot \mathbf{R} = -\frac{d\rho}{ds}$ using the orthogonality of $\mathbf{T}$ and $\mathbf{R}$.

Now we have $\mathbf{B} \cdot \mathbf{R} = -\frac{1}{\tau}\frac{d\rho}{ds}$, which contains a part of the puzzle. Our goal is $\mathbf{R} = -\rho\mathbf{N} - \frac{1}{\tau}\frac{d\rho}{ds}\mathbf{B}$, giving $\mathbf{R}$ as a linear combination of $\mathbf{N}$ and $\mathbf{B}$, so write $\mathbf{R} = \alpha\mathbf{N} + \beta\mathbf{B}$ and solve for α and β by dotting the equation by $\mathbf{N}$ and $\mathbf{B}$ successively, using the orthogonality of $\mathbf{N}$ and $\mathbf{B}$. Thus we find $\mathbf{N} \cdot \mathbf{R} = \alpha$ and $\mathbf{B} \cdot \mathbf{R} = \beta$. We already found that $\mathbf{N} \cdot \mathbf{R} = -\rho$ and $\mathbf{B} \cdot \mathbf{R} = -\frac{1}{\tau}\frac{d\rho}{ds}$, so we obtain finally $\mathbf{R} = -\rho\mathbf{N} - \frac{1}{\tau}\frac{d\rho}{ds}\mathbf{B}$.

Had we not known what we were aiming for, we could have still determined this formula by considering the interesting nature of curves on a sphere: the tangent vector is always perpendicular to the position vector. Then, because the $\mathbf{T}$, $\mathbf{N}$ and $\mathbf{B}$ vectors are mutually perpendicular, $\mathbf{N}$ and $\mathbf{B}$ must line in the same plane as $\mathbf{R}$, so we can write $\mathbf{R}$ as a linear combination $\mathbf{R} = \alpha\mathbf{N} + \beta\mathbf{B}$. We would then proceed to find α and β in the same way we did above.

23. *The Darboux vector* is defined to be

$$\omega = \tau\mathbf{T} + k\mathbf{B}$$

Show that the equation

$$\frac{d\mathbf{U}}{ds} = \omega \times \mathbf{U}$$

is satisfied for $\mathbf{U} = \mathbf{T}$, $\mathbf{N}$, and $\mathbf{B}$. Note the resemblance of this equation to the angular velocity equation (1.22).
Solution: Here are the calculations:

- $\frac{d\mathbf{T}}{ds} = k\mathbf{N}$ and $(\tau\mathbf{T} + k\mathbf{B}) \times \mathbf{T} = 0 + k(\mathbf{B} \times \mathbf{T}) = k\mathbf{N}$.
- $\frac{d\mathbf{N}}{ds} = \tau\mathbf{B} - k\mathbf{T}$ and $(\tau\mathbf{T} + k\mathbf{B}) \times \mathbf{N} = \tau\mathbf{B} - k\mathbf{T}$.
- $\frac{d\mathbf{B}}{ds} = -\tau\mathbf{N}$ and $(\tau\mathbf{T} + k\mathbf{B}) \times \mathbf{B} = -\tau\mathbf{N} + 0 = -\tau\mathbf{N}$.

2.4 Planar Motion in Polar Coordinates

1. Find $\mathbf{v}$ and $\mathbf{a}$ if a particle moves such that

$$r = b(1 - \cos\theta), \qquad \frac{d\theta}{dt} = 4.$$

Solution: We know
$\mathbf{v} = \frac{dr}{dt}\mathbf{u}_r + r\frac{d\theta}{dt}\mathbf{u}_\theta$ and
$\mathbf{a} = \left[\frac{d^2r}{dt^2} - r\left(\frac{d\theta}{dt}\right)^2\right]\mathbf{u}_r + \left[r\frac{d^2\theta}{dt^2} + 2\frac{dr}{dt}\frac{d\theta}{dt}\right]\mathbf{u}_\theta$, so we find

1. $\frac{dr}{dt} = b\sin(\theta)\frac{d\theta}{dt} = 4b\sin\theta$
2. $\frac{d^2r}{dt^2} = 4b\cos(\theta)\frac{d\theta}{dt} = 16b\cos\theta$
3. $\frac{d^2\theta}{dt^2} = 0$

and substitute them into the equations for velocity and acceleration. The results are
$\mathbf{v} = 4b[\sin\theta\,\mathbf{u}_r + (1 - \cos\theta)\mathbf{u}_\theta]$ and
$\mathbf{a} = 16b[(2\cos\theta - 1)\mathbf{u}_r + (2\sin\theta)\mathbf{u}_\theta]$.

3. A particle moves so that its position (r, θ) in polar coordinates is given by

$$r = 2(1 + \sin\theta), \qquad \theta = e^{-t}$$

Find its velocity $\mathbf{v}$ in terms of the vectors $\mathbf{u}_r$ and $\mathbf{u}_\theta$.
Solution: We know
$\mathbf{v} = \frac{dr}{dt}\mathbf{u}_r + r\frac{d\theta}{dt}\mathbf{u}_\theta$, so we find

1. $\frac{d\theta}{dt} = -e^{-t}$
2. $\frac{dr}{dt} = 2\sin(\theta)\frac{d\theta}{dt} = -2e^{-t}\sin\theta$

and substitute them into the equation for velocity to get
$\mathbf{v} = -2e^{-t}[\sin\theta\,\mathbf{u}_r + (1 + \sin\theta)\mathbf{u}_\theta]$.

5. A particle moves along a straight line not passing through the origin. Is $r(d\theta/dt)^2$ nonzero?
Solution: With the conditions given, r can never be zero, so $r(d\theta/dt)^2 = 0$ only if $(d\theta/dt)^2 = 0$. But $\theta = \theta(t)$ is unconstrained in the way it changes in time, except that the particle travels along a line. Unless the velocity is zero, so that the angle of the position vector is constant, this quantity is non-zero.

2.4. PLANAR MOTION IN POLAR COORDINATES

7. A particle moves along the curve

$$r = \frac{1}{1+2\cos\theta}, \qquad \frac{d\theta}{dt} = \frac{1}{r^2}$$

(a) By differentiating the equation $\mathbf{R} = r\mathbf{u}_r$, show that

$$\frac{d\mathbf{R}}{dt} = 2\sin\theta\,\mathbf{u}_r + \frac{1}{r}\mathbf{u}_\theta$$

(b) Find $d^2\mathbf{R}/dt^2$.

Solution:

(a) Write $\frac{d\mathbf{R}}{dt} = \frac{d}{dt}((1+2\cos\theta)^{-1}\mathbf{u}_r) = \frac{(-2\sin\theta\,\frac{d\theta}{dt})}{-(1+2\cos\theta)^2}\mathbf{u}_r + r\frac{d\theta}{dt}\mathbf{u}_\theta$. Substituting $\frac{d\theta}{dt} = \frac{1}{r^2} = (1+2\cos\theta)^2$ and simplifying, we get $\frac{d\mathbf{R}}{dt} = 2\sin\theta\,\mathbf{u}_r + \frac{1}{r}\mathbf{u}_\theta$.

(b) Differentiating $\frac{d\mathbf{R}}{dt} = 2\sin\theta\,\mathbf{u}_r + \frac{1}{r}\mathbf{u}_\theta$ one time, we get $\frac{d^2\mathbf{R}}{dt^2} = 2\cos\theta\,\frac{d\theta}{dt}\mathbf{u}_r + 2\sin\theta\,\mathbf{u}_\theta - \frac{1}{r^2}\mathbf{u}_\theta - \frac{1}{r}\mathbf{u}_r$ which can be simplified to $\frac{d^2\mathbf{R}}{dt^2} = \frac{1}{r}(2\cos\theta - 1)\mathbf{u}_r + (2\sin\theta - \frac{1}{r^2})\mathbf{u}_\theta$.

9. Find the magnitude of the Coriolis acceleration of a particle moving in the xy plane with position given by

$$x = 3t\cos 4\pi t, \qquad y = 3t\sin 4\pi t$$

Solution: The Coriolis acceleration is $2\frac{dr}{dt}\frac{d\theta}{dt}\mathbf{u}_\theta$, so we must find r and θ to begin. Because $x = r\cos\theta$, $y = r\sin\theta$, we know $r = 3t$, $\theta = 4\pi t$, $\frac{dr}{dt} = 3$, $\frac{d\theta}{dt} = 4\pi$, so $2\frac{dr}{dt}\frac{d\theta}{dt}\mathbf{u}_\theta = 24\pi\mathbf{u}_\theta$ and the magnitude of the Coriolis force is 24π.

11. The force $\mathbf{F}$ exerted by a magnetic field $\mathbf{B}$ on a particle carrying a charge q is given by $\mathbf{F} = q(\mathbf{v} \times \mathbf{B})$, where $\mathbf{v}$ is the velocity of the particle. Draw a diagram showing the relative direction of $\mathbf{v}$, $\mathbf{B}$, and $\mathbf{F}$, in some special cases. Under what circumstances will the field exert no force on the particle?
Solution: The illustration shows the situation for the particle moving perpendicular to the magnetic field vector, at an acute angle to, and parallel to the magnetic field vector. When the particle velocity vector is parallel or antiparallel to the magnetic field vector, there will be no force on the particle because the vector product (cross product) is zero.

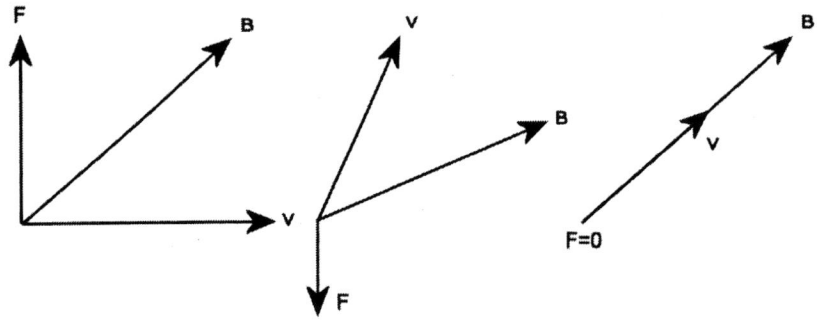

Figure 2.4: The magnetic field vector $\mathbf{B}$, particle velocity vector $\mathbf{v}$ and resulting force vector $\mathbf{F}$ for three situations.

13. An experiment is being designed in which a particle of mass 1 is to exhibit the following planar motion in polar coordinates:

$$r(t) = 1 + t, \qquad \theta(t) = \frac{\pi}{1+t} \text{ for } \quad t \geq 0$$

(a) Determine the position and velocity of this particle at time $t = 1$, illustrating your answer in a diagram.

(b) Find the radial and transverse force $F_r(t)$ and $F_\theta(t)$ needed on the particle to attain the desired motion.

(c) If the forces acting on this particle are removed at t = 1, find its position at t = 5.

Solution: The position at time $t = 1$ is $r = 2$, $\theta = \pi/2$. To compute the velocity and acceleration (and thus the force) we need to compute some derivatives:

$$r = 1 + t \qquad \frac{dr}{dt} = 1 \qquad \frac{d^2r}{dt^2} = 0 \tag{2.1}$$

$$\theta = \frac{\pi}{1+t} \qquad \frac{d\theta}{dt} = \frac{-\pi}{(1+t)^2} \qquad \frac{d^2\theta}{dt^2} = \frac{2\pi}{(1+t)^3} \tag{2.2}$$

The velocity is $\mathbf{v} = \frac{dr}{dt}\mathbf{u}_r + r\frac{d\theta}{dt}\mathbf{u}_\theta$, so our velocity at $t = 1$ is $\mathbf{u}_r - \frac{\pi}{2}\mathbf{u}_\theta$. It is instructive to compute the velocity in cartesian coordinates also. The acceleration is
$\mathbf{a} = \left[\frac{d^2r}{dt^2} - r\left(\frac{d\theta}{dt}\right)^2\right]\mathbf{u}_r + \left[r\frac{d^2\theta}{dt^2} + 2\frac{dr}{dt}\frac{d\theta}{dt}\right]\mathbf{u}_\theta$, so the acceleration (and the force, because the mass is one unit) is
$\mathbf{a} = \left[-r\frac{\pi^2}{(1+t)^4}\right]\mathbf{u}_r + \left[r\frac{2\pi}{(1+t)^3} - 2\frac{\pi}{(1+t)^2}\right]\mathbf{u}_\theta$. The radial and transverse forces are just the components of this vector. Finally, if the forces are removed at $t = 1$, then the position will be determined by the initial position and the velocity at time $t = 1$. The particle will continue in a straight line along the direction of the velocity vector from the point $(0, 2)$ in cartesian coordinates; in other words, the position function is $2\mathbf{j} + t(\pi/2\mathbf{i} + \mathbf{j})$ instantaneously after $t = 1$, and in four more seconds the particle will be at $2\pi\mathbf{i} + 6\mathbf{j}$.

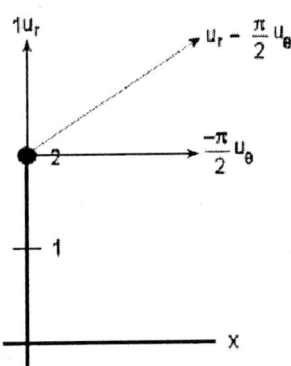

Figure 2.5: The particle at time $t = 1$.

15. A particle moves in a plane with constant angular velocity ω about the origin, but r varies such that the rate of increase of its acceleration is parallel to the position vector $\mathbf{R}$. Show that $d^2r/dt^2 = r\omega^2/3$.

    Solution: The angular velocity $\omega = \frac{d\theta}{dt}$ is constant, so in the velocity and acceleration formulas we can make the replacement $\frac{d\theta}{dt} = k$ and $\frac{d^2\theta}{dt^2} = 0$. Therefore the acceleration equation reads $\mathbf{a} = \left[\frac{d^2r}{dt^2} - rk^2\right]\mathbf{u}_r + \left[2k\frac{dr}{dt}\right]\mathbf{u}_\theta$. We are told that the increase in acceleration is parallel to the position vector, so there is no $\mathbf{u}_\theta$ component of $\frac{d\mathbf{a}}{dt}$. Let us compute the change in acceleration, and equate the coefficient of $\mathbf{u}_\theta$ to 0. After a little work we find

$$\frac{d\mathbf{a}}{dt} = \left[\frac{d^3r}{dt^3} - \frac{dr}{dt}\left(\frac{d\theta}{dt}\right)^2 - 2k\frac{dr}{dt}\frac{d\theta}{dt} - kr\frac{d^2\theta}{dt^2}\right]\mathbf{u}_r + \left[2\frac{d^2r}{dt^2}\frac{d\theta}{dt} + r\frac{d^2\theta}{dt^2} - kr\left(\frac{d\theta}{dt}\right)^2 + k\frac{d^2r}{dt^2}\right]\mathbf{u}_\theta =$$

$$= \left[\frac{d^3r}{dt^3} - k^2\frac{dr}{dt} - 2k^2\frac{dr}{dt}\right]\mathbf{u}_r + \left[3k\frac{d^2r}{dt^2} - k^3r\right]\mathbf{u}_\theta.$$

Now equating the coefficient of $\mathbf{u}_\theta$ to zero, we get the differential equation $3k\frac{d^2r}{dt^2} - k^3r = 0$ or $\frac{d^2r}{dt^2} = r\omega^2/3$.

2.5 Tensor Notation

1. Derive the rule for the derivative of the dot product.
   Solution: The dot product of vectors $\mathbf{v}$ and $\mathbf{w}$ can be written in tensor notation as $\mathbf{v} \cdot \mathbf{w} = \delta_{ij} v_i w_j = v_i w_i = v_j w_j$, where we have used the fact that the substitution tensor δ_{ij} will be zero unless the indices are equal. Assuming the components of each vector are functions of t, we can write the derivative as $\frac{d}{dt}(\mathbf{v} \cdot \mathbf{w}) = \frac{d}{dt}(v_i w_i)$. This is no longer the derivative of a product of vectors: it is a derivative of a product of *components* of vectors indexed by i, so we have a simple product rule $\frac{d}{dt}(v_i w_i) = \frac{dv_i}{dt} w_i + v_i \frac{dw_i}{dt}$. Putting this back into vector notation, we get $\frac{dv_i}{dt} w_i + v_i \frac{dw_i}{dt} = \frac{d\mathbf{v}}{dt} \cdot \mathbf{w} + \mathbf{v} \cdot \frac{d\mathbf{w}}{dt}$.

Chapter 3

Scalar and Vector Fields

3.1 Scalar Fields: Isotimic Surfaces: Gradients

1. Compute **grad** f if

    (a) $f = \sin x + e^{xy} + z$
    (b) $1/|\mathbf{R}|$
    (c) $f = \mathbf{R} \cdot \mathbf{i} \times \mathbf{j}$

    Solution:

    (a) $\mathbf{grad}(\sin x + e^{xy} + z) = \frac{d}{dx}(\sin x + e^{xy} + z)\mathbf{i} + \frac{d}{dy}(\sin x + e^{xy} + z)\mathbf{j} + \frac{d}{dz}(\sin x + e^{xy} + z)\mathbf{k} = (\cos x + ye^{xy})\mathbf{i} + xe^{xy}\mathbf{j} + \mathbf{k}$

    (b) $\mathbf{grad}(1/|\mathbf{R}|) = \mathbf{grad}(\frac{1}{\sqrt{x^2+y^2+z^2}})$
    $= \frac{d}{dx}(\frac{1}{\sqrt{x^2+y^2+z^2}})\mathbf{i} + \frac{d}{dy}(\frac{1}{\sqrt{x^2+y^2+z^2}})\mathbf{j} + \frac{d}{dz}(\frac{1}{\sqrt{x^2+y^2+z^2}})\mathbf{k}$
    $= \frac{d}{dx}(x^2+y^2+z^2)^{-1/2}\mathbf{i} + \frac{d}{dy}(x^2+y^2+z^2)^{-1/2}\mathbf{j} + \frac{d}{dz}(x^2+y^2+z^2)^{-1/2}\mathbf{k}$
    $= -\frac{1}{2}2x(x^2+y^2+z^2)^{-3/2}\mathbf{i} - \frac{1}{2}2y(x^2+y^2+z^2)^{-3/2}\mathbf{j} - \frac{1}{2}2z(x^2+y^2+z^2)^{-3/2}\mathbf{k}$
    $= -x(x^2+y^2+z^2)^{-3/2}\mathbf{i} - y(x^2+y^2+z^2)^{-3/2}\mathbf{j} - z(x^2+y^2+z^2)^{-3/2}\mathbf{k}$
    $= -\frac{x\mathbf{i}+y\mathbf{j}+z\mathbf{k}}{(\sqrt{x^2+y^2+z^2})^3}$
    $= -\mathbf{R}/|\mathbf{R}|^3$

    (c) Because $\mathbf{R} \cdot \mathbf{i} \times \mathbf{j} = \mathbf{R} \cdot (\mathbf{i} \times \mathbf{j}) = (x\mathbf{i}+y\mathbf{j}+z\mathbf{k}) \cdot \begin{vmatrix} \mathbf{i} & \mathbf{j} & \mathbf{k} \\ 1 & 0 & 0 \\ 0 & 1 & 0 \end{vmatrix} = (x\mathbf{i}+y\mathbf{j}+z\mathbf{k}) \cdot \mathbf{k} = z$, we have

    $\mathbf{grad R} \cdot \mathbf{i} \times \mathbf{j} = \frac{d}{dx}(z)\mathbf{i} + \frac{d}{dy}(z)\mathbf{j} + \frac{d}{dz}(z)\mathbf{k} = \mathbf{k}$

3. What can you say about a function whose gradient is everywhere parallel to the y axis?
    Solution: If the function is everywhere parallel to the y axis, then it has no component in the $\mathbf{i}$ or $\mathbf{k}$ direction. The only way this can occur is if the derivatives of f with respect to x and z are zero, which in turn means that x and z do not appear in the function. Thus f is a function depending only on y.

5. Can you describe a scalar whose gradient is $y\mathbf{i}$?
    Solution: There is no such function. Suppose there were such a function $\phi(x,y,z)$. Then its derivative with respect to x must be y, so we can find its form by integrating with respect to x. This integration gives $\phi = xy + f(yz)$, where we have f instead of a constant of integration (we are integrating partial derivatives). Then we can differentiate with respect to y which must yield 0.

3.1. SCALAR FIELDS: ISOTIMIC SURFACES: GRADIENTS

However, we get $\frac{\partial \phi}{\partial y} = x + \frac{\partial f(y,z)}{\partial y} = 0 \Rightarrow x = -\frac{\partial f(y,z)}{\partial y}$, which is impossible because f is a function of y and z only.

7. Given $f(x,y,z) = x^2 + y^2 + z^2$, find the maximum value of $\frac{df}{ds}$ at the point $(3,0,4)$,

   (a) by using the gradient of f.
   (b) by interpreting f geometrically.

   Solution:

   (a) $f = x^2 + y^2 + z^2$ so max of $\frac{df}{ds}$ is $(2x, 2y, 2z)$ which at $(3,0,4)$ is $(6,0,8)$, which is a distance $\sqrt{6^2 + 0^2 + 8^2} = 10$ from the origin.

   (b) The position vector to $(3,0,4)$ is in the direction of greatest change of the function because the surfaces of constant functional value are spherical shells. The change in f with arc length at a point is the position vector itself, so at the point $(3,0,4)$ we have added the vector $3\mathbf{i} + 4\mathbf{k}$ or $6\mathbf{i} + 8\mathbf{k}$, whose magnitude is $\sqrt{6^2 + 0^2 + 8^2} = 10$.

9. Find the derivative of $f(x,y,z) = x + xyz$ at the point $(1,-2,2)$ in the direction of

   (a) $2\mathbf{i} + 2\mathbf{j} - \mathbf{k}$
   (b) $2\mathbf{i} + 2\mathbf{j} + \mathbf{k}$

   Solution:

   (a) **grad** $f = (1+yz)\mathbf{i} + xz\mathbf{j} + xy\mathbf{k}$
   grad $f(1,-2,2) = -3\mathbf{i} + 2\mathbf{j} - 2\mathbf{k}$
   $\mathbf{u} = \frac{2}{3}\mathbf{i} + \frac{2}{3}\mathbf{j} - \frac{1}{3}\mathbf{k}$
   grad $f \cdot \mathbf{u} = \frac{-6+4+2}{3} = 0$

   (b) **grad** $f = (1+yz)\mathbf{i} + xz\mathbf{j} + xy\mathbf{k}$
   grad $f(1,-2,2) = -3\mathbf{i} + 2\mathbf{j} - 2\mathbf{k}$
   $\mathbf{u} = \frac{2}{3}\mathbf{i} + \frac{2}{3}\mathbf{j} + \frac{1}{3}\mathbf{k}$
   grad $f \cdot \mathbf{u} = \frac{-6+4-2}{3} = -4/3$

11. Find the magnitude of the greatest rate of change of $f(x,y,z) = (x^2 + z^2)^3$ at $(1,3,-2)$. Interpret geometrically.
    Solution: $\nabla f = 6x(x^2+z^2)^2\mathbf{i} + 6z(x^2+z^2)^2\mathbf{k}$, and at $(1,3,-2)$ the magnitude is $\sqrt{(6(5)^2)^2 + (6(-2)(5)^2)^2} = 150\sqrt{5}$. Because there is no y in f, the y component of ∇f is 0, and the gradient lies in the x,z plane. Because $(x^2+z^2) = s^2$ is the square of the distance from the y axis, $(x^2+z^2)^3 = s^6$, and $\frac{df}{ds} = 6s^5$.

13. By vector methods, find the point on the curve $x = t, y = t^2, z = 2$ at which the temperature $\phi(x,y,z) = x^2 - 6x + y^2$ takes its minimum value.
    Solution: We could write the temperature function ϕ in terms of the parameterization of the curve $x = t, y = t^2, z = 2$ as $\phi(t) = t^2 - 6t + t^4$. This is the temperature along the curve. We could use calculus to find the value of t that gives a minimum, but our author has requested that we solve the problem by vector methods. The gradient of a function, such as the temperature function, is a vector that points in the direction of the greatest change of the function. The scalar product of a gradient and a tangent to a curve gives the change in the function (in this case temperature) in the direction of the curve. When the change along the curve is zero, we have a max or min. Take the gradient of ϕ: **grad** $\phi = 2(x-3)\mathbf{i} + 2y\mathbf{j}$ and write it in terms of the parameterization of the curve $x = t, y = t^2, z = 2$ to get **grad**$\phi(t) = 2(t-3)\mathbf{i} + 2t^2\mathbf{j}$. Taking the scalar product with the tangent to the curve $\frac{d\mathbf{R}}{dt} = \mathbf{i} + 2t\mathbf{j} + 2\mathbf{k}$, we get $2t - 6 + 4t^3$, which has a zero at $t = 1$. This is our minimum, which we can guarantee by substitution into the original parameterized temperature formula. Thus the point on the curve at which the temperature is a minimum is $\mathbf{i} + \mathbf{j} + 2\mathbf{k}$.

15. Find an equation of the plane tangent to the sphere $x^2 + y^2 + z^2 = 21$ at $(2,4,-1)$.
    Solution: The equation of the plane tangent to the sphere $x^2 + y^2 + z^2 = 21$ at the point $(2,4,-1)$ is found by taking the gradient: $\nabla(x^2 + y^2 + z^2) = 2x\mathbf{i} + 2y\mathbf{j} + 2z\mathbf{k}$, evaluating at $(2,4,-1)$ to get $4\mathbf{i} + 8\mathbf{j} - 2\mathbf{k}$, then writing the equation of the plane $ax + by + cz = ax_0 + by_0 + cz_0$ as $4x + 8y - 2z = 4(2) + 8(4) - 2(-1) = 42$

3.1. SCALAR FIELDS: ISOTIMIC SURFACES: GRADIENTS

17. Find an equation of the plane tangent to the surface $z^2 - xy = 14$ at $(2, 1, 4)$.
Solution: The gradient of the equation of the surface gives a vector normal to the surface, and thus normal to a plane tangent to the surface. The gradient of $z^2 - xy = 14$ is $\nabla(z^2 - xy - 14) = -y\mathbf{i} - x\mathbf{j} + 2z\mathbf{k}$, and at $(2, 1, 4)$ this is $-\mathbf{i} - 2\mathbf{j} + 8\mathbf{k}$. This is normal to the plane $(-\mathbf{i} - 2\mathbf{j} + 8\mathbf{k}) \cdot ((x - 2)\mathbf{i} + (y - 1)\mathbf{j} + (z - 4)\mathbf{k}) = 0$ or $-x - 2y + 8z = 28$.

19. Find the unit vector normal to the plane $3x - y + 2z = 3$,

(a) by the methods of section 1.10.

(b) by the methods of the preceding section.

Solution:

(a) We can read the normal vector right off the plane equation, it is $3\mathbf{i} - \mathbf{j} + 2\mathbf{k}$. Its length is $\sqrt{14}$, so the unit normals are $\mathbf{N} = \pm\frac{1}{\sqrt{14}}(3\mathbf{i} - \mathbf{j} + 2\mathbf{k})$, because the normal can point away from either side of the plane.

(b) We can find the gradient vector, which is normal to the plane by $\nabla(3x - y + 2z - 3) = 3\mathbf{i} - \mathbf{j} + 2\mathbf{k}$, and its length is $\sqrt{14}$, so the unit normals are $\mathbf{N} = \pm\frac{1}{\sqrt{14}}(3\mathbf{i} - \mathbf{j} + 2\mathbf{k})$.

21. Let $T(x, y, z) = x^2 + 2y^2 + 3z^2$, and let S be the isotimic surface: $T = 1$. Find all points (x, y, z) on S that have tangent planes with normals $(1, 1, 1)$.
Solution: (*Note:* We suspect the authors may have updated the notation in the text and missed fixing this. You will sometimes see vectors written $(1, 1, 1)$, $<1, 1, 1>$ and $\mathbf{i} + \mathbf{j} + \mathbf{k}$, but Davis and Snider are very careful to distinguish points (a, b, c) from vectors $a\mathbf{i} + b\mathbf{j} + c\mathbf{k}$. We will interpret $(1, 1, 1)$ as $\mathbf{i} + \mathbf{j} + \mathbf{k}$.) The gradient of T gives a vector normal to the surface, and thus normal to its tangent plane. The gradient is $\nabla T = 2x\mathbf{i} + 4y\mathbf{j} + 6z\mathbf{k}$. We are looking for all the points that satisfy $x^2 + 2y^2 + 3z^2 = 1$ for which $k(2x\mathbf{i} + 4y\mathbf{j} + 6z\mathbf{k}) = (\mathbf{i} + \mathbf{j} + \mathbf{k})$ for some k. That is, $(x, y, z) = (\frac{1}{2k}, \frac{1}{4k}, \frac{1}{6k})$, so that $(\frac{1}{2k})^2 + 2(\frac{1}{4k})^2 + 3(\frac{1}{6k})^2 = 1$ or $\frac{1}{4} + \frac{2}{16} + \frac{3}{36} = k^2$. This has solution $k = \pm\sqrt{\frac{11}{24}}$, so that the points we are looking for are $\pm\left(\frac{1}{2\sqrt{\frac{11}{24}}}, \frac{1}{4\sqrt{\frac{11}{24}}}, \frac{1}{6\sqrt{\frac{11}{24}}}\right) = \pm\sqrt{\frac{6}{11}}\left(1, \frac{1}{2}, \frac{1}{3}\right)$.

23. Find a unit vector tangent to the curve of intersection of the cylinder $x^2 + y^2 = 4$ and the sphere $x^2 + y^2 + z^2 = 9$ at the point $(\sqrt{2}, \sqrt{2}, \sqrt{5})$.
Solution: The intersection of $x^2 + y^2 = 4$ and $x^2 + y^2 + z^2 = 9$ is the set of two circles $x^2 + y^2 = 4$, with $z = \pm\sqrt{5}$. A vector tangent to the curve $x^2 + y^2 - 4 = 0$ at the point $(\sqrt{2}, \sqrt{2}, \sqrt{5})$ is one that is perpendicular to the normal $2x\mathbf{i} + 2y\mathbf{j}$ at that point, $2\sqrt{2}\mathbf{i} + 2\sqrt{2}\mathbf{j}$. One such vector is $-2\sqrt{2}\mathbf{i} + 2\sqrt{2}\mathbf{j}$, and its unit version is $-\frac{\sqrt{2}}{2}\mathbf{i} + \frac{\sqrt{2}}{2}\mathbf{j}$. Of course, we could take the vector in the opposite direction also.

25. At what angle does the line $2x = y = 2z$ intersect the ellipsoid $2x^2 + y^2 + 2z^2 = 8$?
Solution: This should be interpreted to mean "at what angle does a vector in the direction of the line intersect the plane tangent to the ellipsoid." We can most easily calculate this by subtracting the angle between the vector in the direction of the line and the normal to the plane (and thus the ellipsoid to which it is tangent) from $90°$. Solving the equations for the line and the ellipsoid simultaneously, we find the line intersects the ellipsoid at $\pm(1, 2, 1)$. A vector in the direction of the line is $1\mathbf{i} + 2\mathbf{j} + 1\mathbf{k}$ (double that obtained just by writing the line equation in standard form and reading off the vector components), and a normal to the ellipsoid is $\nabla(2x^2 + y^2 + 2z^2 - 8) = 4x\mathbf{i} + 2y\mathbf{j} + 4z\mathbf{k}$. At the point of intersection $(1, 2, 1)$, this is $4\mathbf{i} + 4\mathbf{j} + 4\mathbf{k}$, but we can use $\mathbf{i} + \mathbf{j} + \mathbf{k}$ to make things simpler, as it is in the same direction. Using $\cos\theta = \frac{\mathbf{A}\cdot\mathbf{B}}{|\mathbf{A}||\mathbf{B}|}$, we find $\cos\theta = \frac{(1\mathbf{i}+2\mathbf{j}+1\mathbf{k})\cdot(1\mathbf{i}+1\mathbf{j}+1\mathbf{k})}{|1\mathbf{i}+2\mathbf{j}+1\mathbf{k}||1\mathbf{i}+1\mathbf{j}+1\mathbf{k}|} = \frac{4}{\sqrt{6}\sqrt{3}} = \frac{2\sqrt{2}}{3}$. Because we want $90°$ minus this angle, we can use instead $\sin^{-1}\frac{2\sqrt{2}}{3}$.

27. At what angle does the curve $x = t, y = 2t - t^2, z = 2t^4$ intersect the surface $x^2 + y^3 + 3z^2 = 14$ at the point $(1, 1, 2)$?
Solution: The curve intersects the surface at $(1, 1, 2)$ when $t = 1$. The angle we will figure by using the tangent vector to the curve, which is $\mathbf{i} + (2 - 2t)\mathbf{j} + 8t^3\mathbf{k}$, or, at $t = 1$, $\mathbf{i} + 8\mathbf{k}$. A normal to the surface is $\nabla(x^2 + y^3 + 3z^2 - 14) = 2x\mathbf{i} + 3y^2\mathbf{j} + 6z\mathbf{k}$, which at $(1, 1, 2)$ is $2\mathbf{i} + 3\mathbf{j} + 12\mathbf{k}$. Using $\cos\theta = \frac{\mathbf{A}\cdot\mathbf{B}}{|\mathbf{A}||\mathbf{B}|}$, we find $\cos\theta = \frac{(\mathbf{i}+8\mathbf{k})\cdot(2\mathbf{i}+3\mathbf{j}+12\mathbf{k})}{|\mathbf{i}+8\mathbf{k}||2\mathbf{i}+3\mathbf{j}+12\mathbf{k}|} = \frac{98}{\sqrt{65}\sqrt{157}}$, which gives the angle between the curve and the normal to the surface. The angle we seek is $90° - \cos^{-1}\frac{98}{\sqrt{65}\sqrt{157}}$.

3.1. SCALAR FIELDS: ISOTIMIC SURFACES: GRADIENTS

29. Let S_1 and S_2 be the surfaces with equations

$$\frac{x^2}{a^2} + \frac{y^2}{b^2} + \frac{z^2}{c^2} = 1 \qquad \frac{x^2}{A^2} + \frac{y^2}{B^2} + \frac{z^2}{C^2} = 1.$$

Show that if $a^2B^2 - b^2A^2 = 0$ then the curve of intersection of S_1 and S_2 must be parallel to the xy plane.
Solution: The curve of intersection of the two surfaces will be parallel to the xy plane if the tangent to that curve is, and we can find the tangent to that curve by taking the cross product of the normals to the two surfaces. The normals are $\mathbf{N}_1 = \frac{2x}{a^2}\mathbf{i} + \frac{2y}{b^2}\mathbf{j} + \frac{2z}{c^2}\mathbf{k}$ and $\mathbf{N}_2 = \frac{2x}{A^2}\mathbf{i} + \frac{2y}{B^2}\mathbf{j} + \frac{2z}{C^2}\mathbf{k}$, the tangent to the curve of intersection is $\mathbf{T} = \mathbf{N}_1 \times \mathbf{N}_2 = \begin{vmatrix} \mathbf{i} & \mathbf{j} & \mathbf{k} \\ \frac{2x}{a^2} & \frac{2y}{b^2} & \frac{2z}{c^2} \\ \frac{2x}{A^2} & \frac{2y}{B^2} & \frac{2z}{C^2} \end{vmatrix} =$
$= \left(\frac{4yz}{b^2C^2} - \frac{4yz}{B^2c^2}\right)\mathbf{i} + \left(\frac{4zx}{A^2c^2} - \frac{4zx}{a^2C^2}\right)\mathbf{j} + \left(\frac{4xy}{a^2B^2} - \frac{4xy}{A^2b^2}\right)\mathbf{k}$. If $a^2B^2 - b^2A^2 = 0$, then $a^2B^2 = b^2A^2$ so the component in the $\mathbf{k}$ direction vanishes, and the curve is parallel to the xy plane.

31. Find the point on the sphere $x^2 + y^2 + z^2 = 84$ that is nearest the plane $x + 2y + 4z = 77$.
Solution: The normal to the plane is $1\mathbf{i} + 2\mathbf{j} + 4\mathbf{k}$, and the normal to the sphere is $2x\mathbf{i} + 2y\mathbf{j} + 2z\mathbf{k}$. The shortest distance to the sphere will be that for which the normals are parallel, or $1\mathbf{i} + 2\mathbf{j} + 4\mathbf{k} \times 2x\mathbf{i} + 2y\mathbf{j} + 2z\mathbf{k} = 0$. We find $y = 2x, z = 4x$, and substitute into the equation for the sphere: $x^2 + y^2 + z^2 = 84 = x^2 + (2x)^2 + (4x)^2$ or $x = 2$ so the point on the sphere closest to the plane is $(2, 4, 8)$.

33. What point on the curve $x = t, y = t^2, z = 2$ is closest to the surface $x^2 - 6x + y^2 + 7 = 0$?
Solution: We can complete the square in $x^2 - 6x + y^2 + 7 = 0$ to see that the surface is a cylinder of radius $\sqrt{2}$ centered at $(3, 0, z)$, and because the curve $x = t, y = t^2, z = 2$ lies in the plane $z = 2$, we can do our work in the plane $z = 0$ with the curves $y = x^2$ and $y = \pm\sqrt{2 - (x-3)^2}$. Refer to Figure 3.1, and notice that the outward normal to the circle is in the direction of a line through the point $(3, 0)$. The gradient of $x^2 - y$ is an outward normal $2x\mathbf{i} - \mathbf{j}$ to the curve $y = x^2$ and when the slope $-\frac{1}{2x}$ of the normal to the parabola is equal to the slope of the normal to the circle, the line in this direction, which passes through $(3, 0)$, intersects the parabola in the point closest to the circle. The slope of the normal to the circle is $\frac{x^2}{x-3}$, so equating the two slopes $-\frac{1}{2x} = \frac{x^2}{x-3}$ and solving, we find $x = 1$ and $y = x^2 = 1$. Thus the point on the curve closest to the surface is $(1, 1, 2)$.

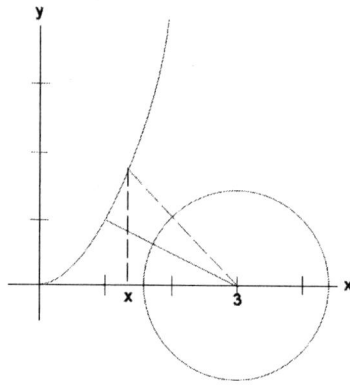

Figure 3.1: The parabola $y = x^2$ and the circle $y = \pm\sqrt{2 - (x-3)^2}$.

35. Given $\phi = \tan^{-1} x + \tan^{-1} y$ and $\Phi = (x+y)/(1-xy)$, show that $\nabla\phi \times \nabla\Phi = 0$. [*Hint:* it is easy if you recognize the formula for $\tan(A+B)$.]
Solution: Consider the sum of angles formula $\tan(\theta_1 \pm \theta_2) = \frac{\tan\theta_1 \pm \tan\theta_2}{1 \mp \tan\theta_1 \tan\theta_2}$; if we make a change of variables $x = \tan A, y = \tan B, z = z$, then $\Phi = (\tan A + \tan B)/(1 - \tan A \tan B) = \tan(A+B)$ and $\phi = A+B$, so $\Phi = \tan\phi$. Now let's revert to the original variables, so that $\phi = \tan^{-1} x + \tan^{-1} y$ and $\Phi = \tan\phi$. Then $\nabla\phi = \frac{1}{x^2+1}\mathbf{i} + \frac{1}{y^2+1}\mathbf{k} + 0\mathbf{j}$ and $\nabla\Phi = (1+\sin^2\phi)\frac{1}{x^2+1}\mathbf{i} + (1+\sin^2\phi)\frac{1}{y^2+1}\mathbf{j} + 0\mathbf{k}$ by the

chain rule. Computing the vector product $\nabla\phi\times\nabla\Phi = \begin{vmatrix} \mathbf{i} & \mathbf{j} & \mathbf{k} \\ \frac{1}{x^2+1} & \frac{1}{y^2+1} & 0 \\ (1+\sin^2\phi)\frac{1}{x^2+1} & (1+\sin^2\phi)\frac{1}{y^2+1} & 0 \end{vmatrix} =$
$\left[(1+\sin^2\phi)\frac{1}{y^2+1}\frac{1}{x^2+1} - (1+\sin^2\phi)\frac{1}{x^2+1}\frac{1}{y^2+1}\right]\mathbf{k} = \mathbf{0}$. More generally, if $f = f(g(x,y))$, then $\nabla g \times \nabla f = f_g g_x g_y - f_g g_y g_x = 0$ by the chain rule as above.

37. Generalize the result of the preceding exercise. *Note:* 36. Given $w = uv$, where u and v are scalar fields, show that $\nabla w \cdot \nabla u \times \nabla v = 0$,

    (a) by direct calculation.
    (b) without calculation.

    Solution:
    (a) Writing out $w = f(u,v)$ and taking the gradient, we have $\nabla w = \frac{\partial w}{\partial u}\nabla u + \frac{\partial w}{\partial v}\nabla v$, which you can see by writing out the derivative by components: $\nabla w = \left(\frac{\partial w}{\partial u}\frac{\partial u}{\partial x} + \frac{\partial w}{\partial v}\frac{\partial v}{\partial x}\right)\mathbf{i} + \left(\frac{\partial w}{\partial u}\frac{\partial u}{\partial y} + \frac{\partial w}{\partial v}\frac{\partial v}{\partial y}\right)\mathbf{j} + \left(\frac{\partial w}{\partial u}\frac{\partial u}{\partial z} + \frac{\partial w}{\partial v}\frac{\partial v}{\partial z}\right)\mathbf{k}$. Thus $\nabla w \cdot \nabla u \times \nabla v = \left(\frac{\partial w}{\partial u}\nabla u + \frac{\partial w}{\partial v}\nabla v\right) \cdot \nabla u \times \nabla v = \frac{\partial w}{\partial u}\nabla u \cdot \nabla u \times \nabla v + \frac{\partial w}{\partial v}\nabla v \cdot \nabla u \times \nabla v = 0$.
    (b) The isotimic surfaces $u = c_1$ and $v = c_2$ intersect in a curve in which both u and v are constant, so if $w = f(u,v)$, w will also be constant along that curve. Because $\nabla u \times \nabla v$ is tangent to that curve, it is perpendicular to w.

3.2 Vector Fields and Flow Lines

1. For the vector field **F** of example 3.12, draw a diagram similar to figure 3.4 showing the values of **F** at the points $(1,0), (0,1), (-1,0), (0,-1), (1,1), (-1,1), (-1,-1), (1,-1)$ and a scattering of other points. Indicate flow lines.
   Solution: Example 3.12 has $\mathbf{F} = -y\mathbf{i} + x\mathbf{j}$. The figure should look something like this:

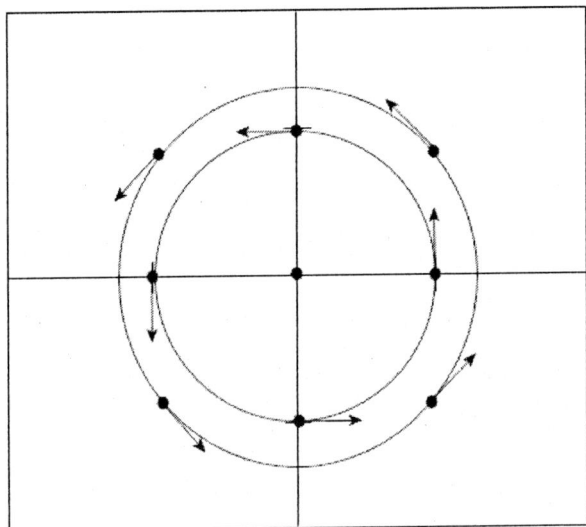

Figure 3.2: To make the figure, we just drew vectors with the components given by the points.

3. With out doing any calculating, describe the flow lines of the vector field $\mathbf{R} = x\mathbf{i} + y\mathbf{j} + z\mathbf{k}$. [*Hint:* If a particle located at (x,y,z) has velocity **R**, in what direction is it moving relative to the origin?]
   Solution: **R** is a vector field pointing in the direction of a position vector, so the flow is radially outward. This is just another way of saying half lines from the origin.

3.3 Divergence

1. Find div **F**, given that $\mathbf{F} = e^{xy}\mathbf{i} + \sin xy\,\mathbf{j} + \cos^2 zx\,\mathbf{k}$.
   Solution: Simply calculate $\frac{\partial}{\partial x}e^{xy} + \frac{\partial}{\partial y}\sin xy + \frac{\partial}{\partial z}\cos^2 zx = ye^{xy} + x\cos xy - 2x\cos zx \sin zx$.

3. Find div **F**, given that $\mathbf{F} = \mathbf{grad}\,\phi$, where $\phi = 3x^2y^3z$.
   Solution: Calculate $\mathbf{F} = 6xy^3z\mathbf{i} + 9x^2y^2z\mathbf{j} + 3x^2y^3\mathbf{k}$, then $\frac{\partial}{\partial x}6xy^3z + \frac{\partial}{\partial y}9x^2y^2z + \frac{\partial}{\partial z}3x^2y^3 = 6y^3z + 18x^2yz$.

5. Show in detail that div $(\phi\mathbf{F}) = \phi\,\text{div}\,\mathbf{F} + \mathbf{F}\cdot\mathbf{grad}\,\phi$.
   Solution: Using the product rule, we can write div $(\phi\mathbf{F}) = \frac{\partial}{\partial x}(\phi F_1) + \frac{\partial}{\partial y}(\phi F_2) + \frac{\partial}{\partial z}(\phi F_3) = \left(\frac{\partial\phi}{\partial x}F_1 + \frac{\partial\phi}{\partial y}F_2 + \frac{\partial\phi}{\partial z}F_3\right) + \phi\left(\frac{\partial F_1}{\partial x} + \frac{\partial F_2}{\partial y} + \frac{\partial F_3}{\partial z}\right) = \mathbf{grad}\phi\cdot\mathbf{F} + \phi\,\text{div}\,\mathbf{F} = \phi\,\text{div}\,\mathbf{F} + \mathbf{F}\cdot\mathbf{grad}\,\phi$.

7. Give an example of a nonconstant field with zero divergence.
   Solution: One example is $-x\mathbf{i} + (y+z)\mathbf{j} + xy\mathbf{k}$ which has divergence $\mathbf{div}(-x\mathbf{i}+(y+z)\mathbf{j}+xy\mathbf{k}) = (\frac{\partial}{\partial x}\mathbf{i}+\frac{\partial}{\partial y}\mathbf{j}+\frac{\partial}{\partial z}\mathbf{k})\cdot(-x\mathbf{i}+(y+z)\mathbf{j}+xy\mathbf{k}) = -1+1+0 = 0$. Another is $<x^2-y, y-2xy, -z>$ which has divergence $\mathbf{div}((x^2-y)\mathbf{i}+(y-2xy)\mathbf{j}-z\mathbf{k}) = (\frac{\partial}{\partial x}\mathbf{i}+\frac{\partial}{\partial y}\mathbf{j}+\frac{\partial}{\partial z}\mathbf{k})\cdot((x^2-y)\mathbf{i}+(y-2xy)\mathbf{j}-z\mathbf{k}) = 2x+1-2x-1 = 0$.

9. Give an example of a field whose divergence depends only on x, is always positive, and increases with increasing x. (Hint: the function e^x is positive for every x.)
   Solution: One example is $<xe^x, yxe^x, 0>$.

11. What can you say about the divergence of the vector field in figure 3.11 at points P, Q and R? Assume no variation if **F** in the z direction and that F_3 is identically zero.
   Solution: There appears to be uniform rotation with no inward or outward component. Because there is no change in the z direction, we need only look at change in the x or y direction. Along the x axis, there does not appear to be any change in the x direction, and along the y axis there does not appear to be any change in the y and similarly for any rotated set or orthogonal axes. Thus, the vector field does not spread apart anywhere, so the divergence is 0.

3.4 Curl

In exercises 1 through 3, find **curl F**.

1. $\mathbf{F} = xy^2\mathbf{i} + xy\mathbf{j} + xy\,\mathbf{k}$
   Solution: $\mathbf{curl}\,\mathbf{F} = (\frac{\partial}{\partial y}(xy) - \frac{\partial}{\partial z}(xy))\mathbf{i} + (\frac{\partial}{\partial z}(xy^2) - \frac{\partial}{\partial x}(xy))\mathbf{j} + (\frac{\partial}{\partial x}(xy) - \frac{\partial}{\partial y}(xy^2))\mathbf{k}$
   $= x\mathbf{i} - y\mathbf{j} + y(1-2x)\,\mathbf{k}$

3. $\mathbf{F} = z^2x\mathbf{i} + y^2z\mathbf{j} - z^2y\mathbf{k}$
   Solution: $\mathbf{curl}\,\mathbf{F} = (\frac{\partial}{\partial y}(-z^2y) - \frac{\partial}{\partial z}(y^2z))\mathbf{i} + (\frac{\partial}{\partial z}(z^2x) - \frac{\partial}{\partial x}(-z^2y))\mathbf{j} + (\frac{\partial}{\partial x}(y^2z) - \frac{\partial}{\partial y}(z^2x))\mathbf{k}$
   $= -(y^2+z^2)\mathbf{i} + (2zx)\mathbf{j}$

5. Draw a rough picture of the vector field $\mathbf{F} = x\mathbf{i} + y\mathbf{j} + z\mathbf{k}$ and, thinking of the paddle wheel interpretation of **curl F**, explain why **curl F** is identically zero in this case.
   Solution: This vector field has vectors whose length is proportional to their distance from the origin. A fluid particle following the flow along the direction of this field has neighboring vectors pointing almost the same direction with the same length, so there is no shear and no local rotation, so no curl.

7. The flow lines of a velocity field **F** are straight lines. Does this imply that **curl F** $= 0$?
   Solution: No. Picture a vector field as a velocity field of a fluid, and imagine that the fluid is full of little neutrally buoyant plastic balls. The curl at each point can be thought of as a vector pointing along the axis of rotation of the balls (using the right hand rule), with the length of the

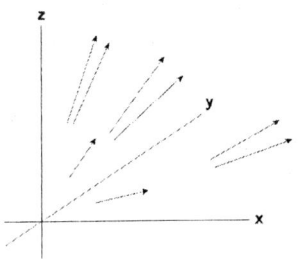

Figure 3.3: A few of the vectors in the radially symmetric vector field.

vector the speed of the rotation. The vectors in a vector field can all be parallel, which corresponds to straight flow lines, while the field has a non-zero curl if there is a shear. If the vectors along one flow line have greater magnitude than those along a neighboring flow line, then we can imagine the fluid is flowing faster in one flow line and slower in the next. This would put a net torque on the balls causing them to spin. As an example, take the vector field $y\mathbf{i}$. This has straight flow lines in the x direction, but the vector field increases in magnitude as y increases. This field has nonzero curl except along the x axis, and the curl is in the $-z$ direction, as it should be from physical considerations.

9. Can you find a vector field whose curl is $y\mathbf{i}$? $x\mathbf{i}$?
    Solution: $\mathbf{F} = y^2/2\,\mathbf{k}$
    curl $\mathbf{F} = (\frac{\partial}{\partial y}(y^2/2) - \frac{\partial}{\partial z}(0))\mathbf{i} + (\frac{\partial}{\partial z}(0) - \frac{\partial}{\partial x}(y^2/2))\mathbf{j} + (\frac{\partial}{\partial x}(0) - \frac{\partial}{\partial y}(0))\mathbf{k} = y\,\mathbf{i}$
    There is no vector field whose curl is $x\mathbf{i}$. The easiest way to see this is to write out the components of **divcurlF** and prove that this must be zero, then take the divergence of $x\mathbf{i}$, which is 1.

11. Let $\mathbf{F}(x,y,z)$ be a vector field defined in all space, and consider an intelligent ant living in the xy plane. Suppose all the ant knows is about $\mathbf{F}$ is its value in the xy plane.

    (a) Can this ant compute **curl F**? Explain briefly.

    (b) Can this ant compute **curl F** $\cdot$ **k**? Explain briefly.

    Solution: An ant in the xy plane cannot measure the component of a vector function out of the plane, so he cannot compute the curl in general. However, the curl dotted with **k** involves only the components of the vector field in the directions of the x and y axes.

13. Show that the x component of **curl F** can be characterized as swirl per unit area for a Δy by Δz rectangle in the y, z plane.
    Solution: Refer to Figure 3.4. The magnitude of the "swirl" can be thought of as the sum of

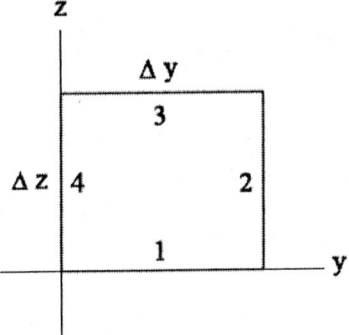

Figure 3.4: A small rectangle of side widths Δy and Δz.

the components of the vector field in the direction of the sides of a geometric figure divided by its area in the limit as the area goes to zero. Alternatively think of the integral of the directional derivative of the field in the direction of a tangent vector to a closed curve divided by the "area" enclosed by the curve (do you see why the "area" here is a vague concept?) in the limit as the length of the curve goes to zero. The direction of the curl is normal to the surface enclosed (again, "normal to the surface" is a vague description because any number of surfaces can be bound by our

3.5. DEL NOTATION

curve). A useful approach is to approximate the geometric figure or curve by a rectangle of side lengths Δy by Δz, say, when we are interested in the x component of the curl. Then the integral of the field over the sides can be approximated by $F_2\Delta y(1) + F_3\Delta z(2) - F_2\Delta y(3) + F_3\Delta z(4) = \Delta y(F_2(1) - F_2(3)) + \Delta z(F_3(2) - F_3(4))$. Now approximate the difference in F_2 between sides 1 and 3 by $\Delta z \frac{\partial F_2}{\partial z}$ and F_3 between sides 2 and 4 by $\Delta y \frac{\partial F_3}{\partial y}$. The "swirl" over the whole area $\Delta y \Delta z$ is approximately $\Delta y \Delta z \frac{\partial F_3}{\partial y} - \Delta y \Delta z \frac{\partial F_2}{\partial z}$, so the swirl per unit area in the x direction is approximately $\frac{\partial F_3}{\partial y} - \frac{\partial F_2}{\partial z}$. In the limit as Δy and Δz go to zero, this becomes exact.

3.5 Del Notation

1. If $f(x, y, z) = x^2 y + z$, what is $f(2, 3, 4)$?
   Solution: $f(2, 3, 4) = 2^2 \cdot 3 + 4 = 16$

3. If $g(t) = t^3$ and $f(x, y, z) = x^2 + y^2 z$, what is $g[f(1, 1, 3)]$?
   Solution: $g[f(1, 1, 3)] = g[1^2 + 1^2 \cdot 3] = g[4] = 4^3 = 64$

5. If **F** is a vector field, is $\nabla \cdot (\nabla \times \mathbf{F})$ a scalar field or a vector field?
   Solution: The curl of **F** is a vector field. The divergence of a vector field is a scalar field. $\nabla \cdot (\nabla \times \mathbf{F})$ is a scalar field.

7. Find $\nabla \cdot \mathbf{R}$ and $\nabla \times \mathbf{R}$ where $\mathbf{R} = x\mathbf{i} + y\mathbf{j} + z\mathbf{k}$.
   Solution: $\nabla \cdot \mathbf{R} = \frac{\partial}{\partial x}(x) + \frac{\partial}{\partial y}(y) + \frac{\partial}{\partial z}(z) = 3$
   $\nabla \times \mathbf{R} = (\frac{\partial}{\partial y}(z) - \frac{\partial}{\partial z}(y))\mathbf{i} + (\frac{\partial}{\partial z}(x) - \frac{\partial}{\partial x}(z))\mathbf{j} + (\frac{\partial}{\partial x}(y) - \frac{\partial}{\partial y}(x))\mathbf{k} = \mathbf{0}$

9. (a) Compute $\nabla \times (\nabla f)$ for the scalar field defined in exercise 8.
   (b) Now do the same thing for for another scalar field f (use any of the scalar fields defined in the preceding problems, or make one up yourself).
   (c) What can you conjecture from this?
   Solution:

   (a) $\nabla(f) = \nabla(xyz + e^{xz}) = (yz + ze^{xz})\mathbf{i} + xz\mathbf{j} + (xy + xe^{xz})\mathbf{k}$
   $\nabla \times (\nabla f) = (\frac{\partial}{\partial y}(xy + xe^{xz}) - \frac{\partial}{\partial z}(xz))\mathbf{i} + (\frac{\partial}{\partial z}(yz + ze^{xz}) - \frac{\partial}{\partial x}(xy + xe^{xz}))\mathbf{j} + (\frac{\partial}{\partial x}(xz) - \frac{\partial}{\partial y}(yz + ze^{xz}))\mathbf{k} = (x - x)\mathbf{i} + ((y + xze^{xz} + e^{xz}) - (y + xze^{xz} + e^{xz}))\mathbf{j} + (z - z)\mathbf{k} = \mathbf{0}$

   (b) $\nabla(f) = \nabla(x^2 y + z) = 2xy\mathbf{i} + x^2\mathbf{j} + \mathbf{k}$
   $\nabla \times (\nabla f) = (\frac{\partial}{\partial y}(1) - \frac{\partial}{\partial z}(x^2))\mathbf{i} + (\frac{\partial}{\partial z}(2xy) - \frac{\partial}{\partial x}(1))\mathbf{j} + (\frac{\partial}{\partial x}(x^2) - \frac{\partial}{\partial y}(2xy))\mathbf{k} = (0 - 0)\mathbf{i} + (0 - 0)\mathbf{j} + (2x - 2x)\mathbf{k} = \mathbf{0}$

   (c) $\nabla \times (\nabla f) = \mathbf{0}$ for all scalar fields.

3.6 The Laplacian

1. Find $\nabla^2 f$, given that $f(x, y, z) = x^5 y z^3$.
   Solution: $\nabla^2 f = \frac{\partial^2}{\partial x^2}(x^5 y z^3) + \frac{\partial^2}{\partial y^2}(x^5 y z^3) + \frac{\partial^2}{\partial z^2}(x^5 y z^3) = 20x^3 y z^3 + 6x^5 y z$

3. Find $\nabla^2 \mathbf{F}$, given that $\mathbf{F} = 3\mathbf{i} + \mathbf{j} - x^2 y^3 z^4 \mathbf{k}$.
   Solution: $\nabla^2 \mathbf{F} = \frac{\partial^2}{\partial x^2}(\mathbf{F}) + \frac{\partial^2}{\partial y^2}(\mathbf{F}) + \frac{\partial^2}{\partial z^2}(\mathbf{F}) = -2y^3 z^4 \mathbf{k} - 6x^2 y z^2 \mathbf{k} - 12x^2 y^3 z^4 \mathbf{k} = -2yz^2(y^2 z^2 + 3x^2 z^2 + 6x^2 y^2)\mathbf{k}$

5. Tell whether each of the following is a scalar field of a vector field, given that f is a scalar field and **F** is a vector field. Two of the expressions are meaningless; determine which two.

(a) ∇f
(b) $\nabla \cdot \mathbf{F}$
(c) $\nabla \times \mathbf{F}$
(d) $\nabla \cdot (\nabla f)$
(e) $\nabla \times (\nabla f)$
(f) $\nabla \times f$
(g) $\nabla^2 \mathbf{F}$
(h) $\nabla \times (\nabla^2 \mathbf{F})$
(i) $\nabla \times (\nabla^2 f)$
(j) $\nabla(\nabla f)$

Solution:

(a) Vector field.

(b) Scalar field.

(c) Vector field.

(d) Scalar field.

(e) Zero vector field.

(f) Meaningless since the curl of a scalar field does not exist.

(g) Vector field.

(h) Vector field.

(i) Meaningless since $\nabla^2 f$ is a scalar field.

(j) Vector field.

7. Given $f(x,y,z) = 2x^2 + y$ and $\mathbf{R} = x\mathbf{i} + y\mathbf{j} + z\mathbf{k}$ find

(a) ∇f
(b) $\nabla \cdot \mathbf{R}$
(c) $\nabla^2 f$
(d) $\nabla \times (f\mathbf{R})$

Solution:

(a) $\nabla f = \frac{\partial f}{\partial x}\mathbf{i} + \frac{\partial f}{\partial y}\mathbf{j} + \frac{\partial f}{\partial z}\mathbf{k} = 4x\mathbf{i} + \mathbf{j}$

(b) $\nabla \cdot \mathbf{R} = \frac{\partial R_1}{\partial x} + \frac{\partial R_2}{\partial y} + \frac{\partial R_3}{\partial z} = 3$

(c) $\nabla^2 f = \frac{\partial^2 f}{\partial x^2} + \frac{\partial^2 f}{\partial y^2} + \frac{\partial^2 f}{\partial z^2} = 4$

(d) $\nabla \times (f\mathbf{R}) = \nabla \times ((2x^2+y)x\mathbf{i} + (2x^2+y)y\mathbf{j} + (2x^2+y)z\mathbf{k}) = \begin{vmatrix} \mathbf{i} & \mathbf{j} & \mathbf{k} \\ \partial_x & \partial_y & \partial_z \\ (2x^2+y)x & (2x^2+y)y & (2x^2+y)z \end{vmatrix} =$

$z\mathbf{i} - 4xz\mathbf{j} + (4xy - x)\mathbf{k}$

3.7 Dyadics: Taylor Polynomials

1. Verify that the zeroth, first, and second partial derivatives of f and its second-degree Taylor polynomial agree at the base point $\mathbf{R}_0$.
Solution: The Taylor polynomial of second-degree for a function of three variables is

$f(x,y,z) = f(x_0,y_0,z_0) + \frac{1}{1!}(f_x(x_0,y_0,z_0)(x-x_0) + f_y(x_0,y_0,z_0)(y-y_0) + f_z(x_0,y_0,z_0)(z-z_0)) +$
$\frac{1}{2!}(f_{xx}(x_0,y_0,z_0)(x-x_0)^2 + f_{yy}(x_0,y_0,z_0)(y-y_0)^2 + f_{zz}(x_0,y_0,z_0)(z-z_0)^2 +$
$2f_{yz}(x_0,y_0,z_0)(y-y_0)(z-z_0) + 2f_{xz}(x_0,y_0,z_0)(x-x_0)(z-z_0) +$
$2f_{xy}(x_0,y_0,z_0)(x-x_0)(y-y_0))$

If we put in $x = x_0$, $y = y_0$ and $z = z_0$ then we see that the zeroth derivative of f agrees with the Taylor polynomial

$f(x_0,y_0,z_0) = f(x_0,y_0,z_0) + \frac{1}{1!}(f_x(x_0,y_0,z_0)(0) + ...) + \frac{1}{2!}(f_{xx}(x_0,y_0,z_0)(0)^2 + ...)$
$= f(x_0,y_0,z_0)$

3.7. DYADICS: TAYLOR POLYNOMIALS

We can now apply the derivative operator $\mathbf{u}\nabla$ to the original polynomial to get f'. The derivative operator is more easily understood in its expanded form, $\mathbf{u}\nabla = u_1\frac{d}{dx} + u_2\frac{d}{dy} + u_3\frac{d}{dz}$.

$$\begin{aligned}\mathbf{u}\nabla f(x,y,z) = \ &u_1 f_x(x_0,y_0,z_0) + u_2 f_y(x_0,y_0,z_0) + u_3 f_z(x_0,y_0,z_0) + u_1 f_{xx}(x_0,y_0,z_0)(x-x_0) +\\ &u_2 f_{yy}(x_0,y_0,z_0)(y-y_0) + u_3 f_{zz}(x_0,y_0,z_0)(z-z_0) + u_2 f_{yz}(x_0,y_0,z_0)(z-z_0) +\\ &u_3 f_{yz}(x_0,y_0,z_0)(z-z_0) + u_1 f_{xz}(x_0,y_0,z_0)(z-z_0) + u_3 f_{xz}(x_0,y_0,z_0)(x-x_0) +\\ &u_1 f_{xy}(x_0,y_0,z_0)(y-y_0) + u_2 f_{xy}(x_0,y_0,z_0)(x-x_0)\end{aligned}$$

If we again put in $x = x_0$, $y = y_0$ and $z = z_0$ then we see that the first derivative of f agrees with the Taylor polynomial

$$\begin{aligned}f'(x_0,y_0,z_0) = \ &u_1 f_x(x_0,y_0,z_0) + u_2 f_y(x_0,y_0,z_0) + u_3 f_z(x_0,y_0,z_0) + u_1 f_{xx}(x_0,y_0,z_0)(0) +\\ &u_2 f_{yy}(x_0,y_0,z_0)(0) + u_3 f_{zz}(x_0,y_0,z_0)(0) + u_2 f_{yz}(x_0,y_0,z_0)(0) +\\ &u_3 f_{yz}(x_0,y_0,z_0)(0) + u_1 f_{xz}(x_0,y_0,z_0)(0) + u_3 f_{xz}(x_0,y_0,z_0)(0) +\\ &u_1 f_{xy}(x_0,y_0,z_0)(0) + u_2 f_{xy}(x_0,y_0,z_0)(0)\\ = \ &u_1 f_x(x_0,y_0,z_0) + u_2 f_y(x_0,y_0,z_0) + u_3 f_z(x_0,y_0,z_0)\end{aligned}$$

We can now generalize the second-degree Taylor polynomial to functions of any number of variables as,
$$f(\mathbf{R}) = f(\mathbf{R}_0) + (\mathbf{R}-\mathbf{R}_0)\cdot\nabla f(\mathbf{R}_0) + \tfrac{1}{2}(\mathbf{R}-\mathbf{R}_0)\cdot\nabla\nabla f(\mathbf{R}_0)\cdot(\mathbf{R}-\mathbf{R}_0)$$
or
$$f(\mathbf{R}) = \sum_{m=0}^{n}\tfrac{1}{m!}\left[(\mathbf{R}-\mathbf{R}_0)\cdot(\nabla)^m f(\mathbf{R}_0)\right]$$
The zeroth derivative of f evaluated at $\mathbf{R}_0$ is
$$\begin{aligned}f(\mathbf{R}_0) &= \sum_{m=0}^{n}\tfrac{1}{m!}\left[(\mathbf{R}_0-\mathbf{R}_0)\cdot(\nabla)^m f(\mathbf{R}_0)\right]\\ &= f(\mathbf{R}_0)\end{aligned}$$
The first derivative of f evaluated at $\mathbf{R}_0$ is
$$\begin{aligned}f'(\mathbf{R}) &= \sum_{m=1}^{n}\tfrac{1}{(m-1)!}\left[(\mathbf{R}-\mathbf{R}_0)\cdot(\nabla)^{(m-1)} f'(\mathbf{R}_0)\right]\\ f'(\mathbf{R}_0) &= \sum_{m=1}^{n}\tfrac{1}{(m-1)!}\left[(\mathbf{R}_0-\mathbf{R}_0)\cdot(\nabla)^{(m-1)} f'(\mathbf{R}_0)\right]\\ &= f'(\mathbf{R}_0)\end{aligned}$$
And finally, the second derivative of f is
$$\begin{aligned}f''(\mathbf{R}) &= \sum_{m=2}^{n}\tfrac{1}{(m-2)!}\left[(\mathbf{R}-\mathbf{R}_0)\cdot(\nabla)^{(m-2)} f''(\mathbf{R}_0)\right]\\ f''(\mathbf{R}_0) &= \sum_{m=2}^{n}\tfrac{1}{(m-2)!}\left[(\mathbf{R}_0-\mathbf{R}_0)\cdot(\nabla)^{(m-2)} f''(\mathbf{R}_0)\right]\\ &= f''(\mathbf{R}_0)\end{aligned}$$
So, the zeroth, first, and second partial derivatives of f and its second-degree Taylor polynomial agree at the base point $\mathbf{R}_0$.

3. Work out the second order Taylor polynomial for a function of two variables $f(x,y)$.
Solution: The general equation for the third order Taylor polynomial is
$$f(\mathbf{R}) \simeq f(\mathbf{R}_0) + w_i\tfrac{\partial f}{\partial R_i}\big|_{\mathbf{R}_0} + \tfrac{1}{2!}w_i\tfrac{\partial^2 f}{\partial R_i\partial R_j}\big|_{\mathbf{R}_0} w_j + \tfrac{1}{3!}w_i\tfrac{\partial^3 f}{\partial R_i\partial R_j\partial R_k}\big|_{\mathbf{R}_0} w_j w_k$$
where $w_i = \mathbf{R}-\mathbf{R}_0$ so for a function of two variables P_3 will be
$$f(x,y) \simeq f(x_0,y_0) + \begin{bmatrix} x-x_0 & y-y_0 \end{bmatrix}\begin{bmatrix} f_x(x_0,y_0) & f_y(x_0,y_0) \end{bmatrix}$$
$$+ \tfrac{1}{2!}\begin{bmatrix} x-x_0 & y-y_0 \end{bmatrix}\begin{bmatrix} f_{xx}(x_0,y_0) & f_{xy}(x_0,y_0) \\ f_{yx}(x_0,y_0) & f_{yy}(x_0,y_0) \end{bmatrix}\begin{bmatrix} x-x_0 & y-y_0 \end{bmatrix} + \tfrac{1}{3!}w_i\tfrac{\partial^3 f}{\partial R_i\partial R_j\partial R_k}\big|_{\mathbf{R}_0} w_j w_k$$

5. Show that the difference between $f(0,0,0)$ and its average over a small sphere of radius R is given approximately by $-(R^2/10)\nabla^2 f(0,0,0)$. (*Hint:* The average value of x^2 over the sphere is $R^2/10$.
Solution: Using the same methods as example 3.25, we can see that the Taylor polynomial will be $f(0,0,0) + xf_x + yf_y + zf_z + \tfrac{x^2}{2}f_{xx} + \tfrac{y^2}{2}f_{yy} + \tfrac{z^2}{2}f_{zz} + xyf_{xy} + xzf_{xz} + yzf_{yz}$
As in the example, the first term is constant and the average of x, y, z, xy, xz, yz are zero by symmetry. Also because of symmetry we only need to look at the average value of x^2.
The average of x^2 is $\frac{1}{volume}\iiint x^2$ Volume Element which is $\frac{3}{4\pi R^3}\int_0^R\int_0^\pi\int_0^{2\pi}(r\sin\phi\cos\theta)^2 r^2\sin\phi\, dr d\phi d\theta = \frac{3}{4\pi R^3}\frac{R^5}{5}\pi\int_0^\pi\sin^3\phi d\phi = \frac{3}{4\pi R^3}\frac{R^5}{5}\pi = \frac{3}{4}\frac{R^2}{5}$ Therefore, the difference between $f(0,0,0)$ and its average value over a small sphere of radius R is given by
$$\begin{aligned}f(0,0,0) - f_{av} &= -\tfrac{R^2}{10}(f_{xx}+f_{yy}+f_{zz})\\ &= -\tfrac{R^2}{10}\nabla^2 f(0,0,0)\end{aligned}$$

7. Show that $\nabla^2 f \leq 0$ at a local maximum of f, and $\nabla^2 f \geq 0$ at a minimum. (*Hint:* consider f as a function of x, y, and z separately.) Note that this statement agrees with the interpretation of $\nabla^2 f$ as a measure of the difference of the averaged value of f and its local value.
Solution: The laplacian of f written out is $\frac{\partial^2 f}{\partial x^2} + \frac{\partial^2 f}{\partial y^2} + \frac{\partial^2 f}{\partial z^2}$, whose separate components are the

second derivatives of f with respect to x, y, and z. Geometrically, we can think of f as a surface in four space, so if f has a local maximum at a point p, then in a small region around p, there is no curve in the surface that has a positive directional derivative at that point. If there was, then just a little way along that direction must be a value higher for f. In particular, there can be no positive directional derivative at p in any of the x, y or z directions, so the change in the derivatives of curves in f in the x, y, or z directions (the x, y and z components of the laplacian of f) must be zero or negative.

3.8 Vector Identities

1. Verify eqns. (3.27) and (3.28).
   Solution: Equation (3.27) is $\nabla(\phi_1 \phi_2) = \phi_1 \nabla \phi_2 + \phi_2 \nabla \phi_1$. Using ∂_i to indicate differentiation with respect to the ith coordinate direction $x_i = x, y, z$ for $i = 1, 2, 3$, and with the sum $\partial_i \phi \mathbf{e}_i = \sum_{i=1}^{3} \partial_i \phi \mathbf{e}_i = \phi_x \mathbf{i} + \phi_y \mathbf{j} + \phi_z \mathbf{k}$ implied, we write
   $\nabla(\phi_1 \phi_2) = \partial_i(\phi_1 \phi_2) \mathbf{e}_i = ((\partial_i \phi_1)\phi_2 + \phi_1(\partial_i \phi_2))\mathbf{e}_i = \phi_1 \nabla \phi_2 + \phi_2 \nabla \phi_1$
   Equation (3.28) is $\nabla \cdot \phi \mathbf{F} = \phi \nabla \cdot \mathbf{F} + \mathbf{F} \cdot \nabla \phi$, so
   $\nabla \cdot \phi \mathbf{F} = \frac{\partial}{\partial x_i}(\phi F_i) = \frac{\partial F_i}{\partial x_i}\phi + \frac{\partial \phi}{\partial x_i} F_i = \phi \nabla \cdot \mathbf{F} + \mathbf{F} \cdot \nabla \phi$

3. Verify (3.40) through (3.42).
   Solution:

   (a) Identity 3.40 is $\nabla \times \nabla(\phi) = 0$. Write $\nabla(\phi) = \frac{\partial \phi}{\partial x}\mathbf{i} + \frac{\partial \phi}{\partial y}\mathbf{j} + \frac{\partial \phi}{\partial z}\mathbf{k}$ and evaluate the symbolic
   determinant $\begin{vmatrix} \mathbf{i} & \mathbf{j} & \mathbf{k} \\ \frac{\partial}{\partial x} & \frac{\partial}{\partial y} & \frac{\partial}{\partial z} \\ \frac{\partial \phi}{\partial x} & \frac{\partial \phi}{\partial y} & \frac{\partial \phi}{\partial z} \end{vmatrix} = \left(\frac{\partial^2 \phi}{\partial y \partial z} - \frac{\partial^2 \phi}{\partial z \partial y} \right) \mathbf{i} + \left(\frac{\partial^2 \phi}{\partial z \partial x} - \frac{\partial^2 \phi}{\partial x \partial z} \right) \mathbf{j} + \left(\frac{\partial^2 \phi}{\partial x \partial y} - \frac{\partial^2 \phi}{\partial y \partial x} \right) \mathbf{k} = \mathbf{0}$
   because of the equality of partial derivatives in cartesian coordinates.

   (b) Identity 3.41 is $\nabla \cdot (\nabla \times \mathbf{F}) = 0$, so let's compute $\nabla \times \mathbf{F} = \begin{vmatrix} \mathbf{i} & \mathbf{j} & \mathbf{k} \\ \frac{\partial}{\partial x} & \frac{\partial}{\partial y} & \frac{\partial}{\partial z} \\ F_1 & F_2 & F_3 \end{vmatrix} = \left(\frac{\partial F_3}{\partial y} - \frac{\partial F_2}{\partial z} \right) \mathbf{i} +$
   $\left(\frac{\partial F_1}{\partial z} - \frac{\partial F_3}{\partial x} \right) \mathbf{j} + \left(\frac{\partial F_2}{\partial x} - \frac{\partial F_1}{\partial y} \right) \mathbf{k}$. Finally, computing the divergence $\nabla \cdot (\nabla \times \mathbf{F}) = \frac{\partial}{\partial x}\left(\frac{\partial F_3}{\partial y} - \frac{\partial F_2}{\partial z} \right) +$
   $\frac{\partial}{\partial y}\left(\frac{\partial F_1}{\partial z} - \frac{\partial F_3}{\partial x} \right) + \frac{\partial}{\partial z}\left(\frac{\partial F_2}{\partial x} - \frac{\partial F_1}{\partial y} \right) = \frac{\partial^2 F_3}{\partial x \partial y} - \frac{\partial^2 F_3}{\partial y \partial x} + \frac{\partial^2 F_2}{\partial x \partial z} - \frac{\partial^2 F_2}{\partial z \partial x} + \frac{\partial^2 F_1}{\partial z \partial y} - \frac{\partial^2 F_1}{\partial y \partial z} = 0$

   (c) Identity 3.42 is $\nabla \cdot (\nabla \phi_1 \times \nabla \phi_2) = 0$. We know that $\nabla \phi_1 = \frac{\partial \phi_1}{\partial x}\mathbf{i} + \frac{\partial \phi_1}{\partial y}\mathbf{j} + \frac{\partial \phi_1}{\partial z}\mathbf{k}$ and $\nabla \phi_2 = \frac{\partial \phi_2}{\partial x}\mathbf{i} + \frac{\partial \phi_2}{\partial y}\mathbf{j} + \frac{\partial \phi_2}{\partial z}\mathbf{k}$, so placing these in the determinant notation for the triple scalar product
   $\nabla \phi_1 \times \nabla \phi_2 = \begin{vmatrix} \frac{\partial}{\partial x} & \frac{\partial}{\partial y} & \frac{\partial}{\partial z} \\ \frac{\partial \phi_1}{\partial x} & \frac{\partial \phi_1}{\partial y} & \frac{\partial \phi_1}{\partial z} \\ \frac{\partial \phi_2}{\partial x} & \frac{\partial \phi_2}{\partial y} & \frac{\partial \phi_2}{\partial z} \end{vmatrix}$. Then we have $\nabla \cdot (\nabla \phi_1 \times \nabla \phi_2) = \frac{\partial}{\partial x}\left(\frac{\partial \phi_1}{\partial y}\frac{\partial \phi_2}{\partial z} - \frac{\partial \phi_1}{\partial z}\frac{\partial \phi_2}{\partial y} \right) +$
   $\frac{\partial}{\partial x}\left(\frac{\partial \phi_1}{\partial z}\frac{\partial \phi_2}{\partial x} - \frac{\partial \phi_1}{\partial x}\frac{\partial \phi_2}{\partial z} \right) + \frac{\partial}{\partial x}\left(\frac{\partial \phi_1}{\partial x}\frac{\partial \phi_2}{\partial y} - \frac{\partial \phi_1}{\partial y}\frac{\partial \phi_2}{\partial x} \right) = 0$

5. Why is the following "identity" obviously not valid? (*Hint:* Check the symmetry.) $\nabla \cdot (\mathbf{F} \times \mathbf{G}) = \mathbf{G} \cdot (\nabla \times \mathbf{F}) + \mathbf{F} \cdot (\nabla \times \mathbf{G})$.
   Solution: Exchanging $\mathbf{F}$ and $\mathbf{G}$ changes the sign of the left hand side but not the right hand side.

7. If $\mathbf{V}(\mathbf{R})$ can be expressed as $\mathbf{V}(\mathbf{R}) = \mathbf{A}f(\mathbf{R} \cdot \mathbf{B})$, where $\mathbf{A}$ and $\mathbf{B}$ are constant, prove that **curl V** is perpendicular to both $\mathbf{A}$ and $\mathbf{B}$.
   Solution: Using vector identity 3.29 and the fact that the curl of a constant vector is zero, we get $\nabla \times \mathbf{V}(\mathbf{R}) = \nabla f(\mathbf{R} \cdot \mathbf{B}) \times \mathbf{A}$. Now, we can compute the gradient of an unknown function of $\mathbf{R} \cdot \mathbf{B} = b_1 x \mathbf{i} + b_2 y \mathbf{j} + b_3 z \mathbf{k}$ using the chain rule to get $\nabla f = f' \nabla(\mathbf{R} \cdot \mathbf{B}) = b_1 f'(\mathbf{R} \cdot \mathbf{B})\mathbf{i} + b_2 f'(\mathbf{R} \cdot \mathbf{B})\mathbf{j} + b_3 f'(\mathbf{R} \cdot \mathbf{B})\mathbf{k} = f'\mathbf{B}$. Therefore, $\nabla \times \mathbf{V}(\mathbf{R}) = f' \mathbf{B} \times \mathbf{A}$, which is perpendicular to both $\mathbf{A}$ and $\mathbf{B}$.

9. Evaluate $\mathbf{A} \cdot \nabla \mathbf{R} + \nabla(\mathbf{A} \cdot \mathbf{R}) + \mathbf{A} \times \nabla \mathbf{R}$, where $\mathbf{A}$ is a constant vector field.
   Solution: Because $\nabla \times \mathbf{R} = 0$, and by 3.35 in the table $\nabla(\mathbf{A} \cdot \mathbf{R} = \mathbf{A})$ we need only calculate $(\mathbf{A} \cdot \nabla)\mathbf{A} = \mathbf{A}$, so the answer is $2\mathbf{A}$.

3.9. TENSOR NOTATION

11. Evaluate $\nabla \left(\mathbf{A} \cdot \nabla \frac{1}{R}\right) + \nabla \times \left(\mathbf{A} \times \nabla \frac{1}{R}\right)$.
  Solution: Using vector identities

(3.39) $\qquad \nabla(\mathbf{F} \cdot \mathbf{G}) = (\mathbf{F} \cdot \nabla)\mathbf{G} + (\mathbf{G} \cdot \nabla)\mathbf{F} + \mathbf{F} \times (\nabla \times \mathbf{G}) + \mathbf{G} \times (\nabla \times \mathbf{F})$

and

(3.37) $\qquad \nabla \times (\mathbf{F} \times \mathbf{G}) = (\mathbf{G} \cdot \nabla)\mathbf{F} - (\mathbf{F} \cdot \nabla)\mathbf{G} + (\nabla \cdot \mathbf{G})\mathbf{F} - (\nabla \cdot \mathbf{F})\mathbf{G}$

we can rewrite $\nabla \left(\mathbf{A} \cdot \nabla \frac{1}{R}\right) + \nabla \times \left(\mathbf{A} \times \nabla \frac{1}{R}\right)$ as

$$(\mathbf{A} \cdot \nabla)\nabla\frac{1}{R} + (\nabla\frac{1}{R} \cdot \nabla)\mathbf{A} + \mathbf{A} \times (\nabla \times \nabla\frac{1}{R}) + \nabla\frac{1}{R} \times (\nabla \times \mathbf{A}) +$$
$$+ (\nabla\frac{1}{R} \cdot \nabla)\mathbf{A} - (\mathbf{A} \cdot \nabla)\nabla\frac{1}{R} + (\nabla \cdot \nabla\frac{1}{R})\mathbf{A} - (\nabla \cdot \mathbf{A})\nabla\frac{1}{R}.$$

The first term in the first line and the second term in the second line cancel, the second term in the first line and the first term in the second line are zero (a differential operator is being applied to a constant vector), and because $\nabla \times \mathbf{A} = \mathbf{0}$ and $\nabla \times \nabla\frac{1}{R} = \mathbf{0}$ (the curl of a gradient is zero), the remainder of the first line is zero. The last term in the second line vanishes because $\nabla \cdot \mathbf{A} = 0$, so we are left with

$$\nabla \left(\mathbf{A} \cdot \nabla\frac{1}{R}\right) + \nabla \times \left(\mathbf{A} \times \nabla\frac{1}{R}\right) = (\nabla \cdot \nabla\frac{1}{R})\mathbf{A}.$$

The gradient of $\frac{1}{|\mathbf{R}|}$ is $\frac{-\mathbf{R}}{|\mathbf{R}|^3} = -x\left(x^2+y^2+z^2\right)^{3/2}\mathbf{i} - y\left(x^2+y^2+z^2\right)^{3/2}\mathbf{j} - z\left(x^2+y^2+z^2\right)^{3/2}\mathbf{k}$, and the divergence of this quantity is

$$\frac{3x^2}{(x^2+y^2+z^2)^{5/2}} + \frac{3y^2}{(x^2+y^2+z^2)^{5/2}} + \frac{3z^2}{(x^2+y^2+z^2)^{5/2}} - \frac{3}{(x^2+y^2+z^2)^{3/2}}.$$

Putting the last term over the same denominator as the first three gives

$$\frac{3x^2}{(x^2+y^2+z^2)^{5/2}} + \frac{3y^2}{(x^2+y^2+z^2)^{5/2}} + \frac{3z^2}{(x^2+y^2+z^2)^{5/2}} - \frac{3(x^2+y^2+z^2)}{(x^2+y^2+z^2)^{5/2}} = 0.$$

13. Derive the identity $\nabla|\mathbf{F}|^2 = 2\mathbf{F} \cdot \nabla\mathbf{F} + 2\mathbf{F} \times (\nabla \times \mathbf{F})$.
  Solution: Using the vector identity 3.39 $\nabla(\mathbf{F} \cdot \mathbf{G}) = (\mathbf{F} \cdot \nabla)\mathbf{G} + (\mathbf{G} \cdot \nabla)\mathbf{F} + \mathbf{F} \times (\nabla \times \mathbf{G}) + \mathbf{G} \times (\nabla \times \mathbf{F})$ and writing $|\mathbf{F}|^2 = \mathbf{F} \cdot \mathbf{F}$, we see immediately that $\nabla|\mathbf{F}|^2 = \mathbf{F} \cdot \mathbf{F} = \mathbf{F} \cdot \nabla\mathbf{F} + \mathbf{F} \cdot \nabla\mathbf{F} + \mathbf{F} \times (\nabla \times \mathbf{F}) + \mathbf{F} \times (\nabla \times \mathbf{F}) = 2\mathbf{F} \cdot \nabla\mathbf{F} + 2\mathbf{F} \times (\nabla \times \mathbf{F})$.

3.9 Tensor Notation

Using tensor notation, prove the following identities:

1. $\nabla \cdot \phi\mathbf{F} = \phi\nabla \cdot \mathbf{F} + \mathbf{F} \cdot \nabla\phi$
  Solution: $\nabla \cdot (\phi\mathbf{F}) = \partial_i(\phi\mathbf{F})_i = (\partial_i\phi)\mathbf{F} + \phi\partial_i F_i = \nabla\phi \cdot \mathbf{F} + \phi\nabla \cdot \mathbf{F}$.

3. $\nabla \times \mathbf{R} = 0$
  Solution: $\nabla \times \mathbf{R} = \epsilon_{ijk}\partial_j R_k$, and because $\partial_j R_k = \partial_k R_k$ because $\mathbf{R} = (x,y,z)$, we have $\epsilon_{ijk}\partial_k R_k = 0$ by the properties of the permutation tensor.

5. $\nabla \times (\nabla \times \mathbf{F}) = \nabla(\nabla \cdot \mathbf{F}) - \nabla^2\mathbf{F}$
  Solution:

$$\nabla \times (\nabla \times \mathbf{F}) = \epsilon_{ijk}\partial_j(\nabla \times \mathbf{F})_k = \epsilon_{ijk}\partial_j\epsilon_{klm}\partial_l F_m = \epsilon_{ijk}\epsilon_{klm}\partial_j\partial_l F_m = \epsilon_{ijk}\epsilon_{lmk}\partial_j\partial_l F_m = (\delta_{il}\delta_{jm} - \delta_{im}\delta_{jl})\partial_j\partial_l F_m$$
$$= \delta_{il}\delta_{jm}\partial_j\partial_l F_m - \delta_{im}\delta_{jl}\partial_j\partial_l F_m = \delta_{il}\delta_{jj}\partial_j\partial_l F_j - \delta_{im}\delta_{ll}\partial_l\partial_l F_m = \delta_{il}\partial_j\partial_l F_j - \delta_{im}\partial_l\partial_l F_m = \delta_{il}\partial_l\partial_j F_j - \delta_{im}\partial_l\partial_l F_m$$
$$= \delta_{ll}\partial_l\nabla\mathbf{F} - \delta_{mm}\partial_l\partial_l F_m = \partial_l(\nabla \cdot \mathbf{F}) - \partial_l\partial_l\mathbf{F} = \nabla(\nabla \cdot \mathbf{F}) - \nabla^2\mathbf{F}$$

3.10. CYLINDRICAL AND SPHERICAL COORDINATES

7. $\nabla \cdot (\nabla \times \mathbf{F}) = 0$
   Solution:
$$\nabla \cdot (\nabla \times \mathbf{F}) = \partial_i \epsilon_{ijk} \partial_j F_k = \partial_k \epsilon_{ijk} \partial_i F_k = -\partial_k \epsilon_{jik} \partial_i F_k = -\nabla \cdot (\nabla \times \mathbf{F}),$$

so
$$\nabla \cdot (\nabla \times \mathbf{F}) = 0$$

3.10 Cylindrical and Spherical Coordinates

1. Derive the equations of transformation between cylindrical and spherical coordinates.
   Solution: In cylindrical coordinates a point is specified by a height z, a radius r, and an angle θ and in spherical coordinates a point is specified by two angles θ and ϕ and a length ρ. To transform cylindrical information into spherical, we can easily see that θ remains the same. The distance r to the point is $\sqrt{z^2 + \rho^2}$ since ρ, z, r form a right triangle. Finally, ϕ is the $\arctan \rho/z$ since it is the angle measured from the positive z axis.
   $r = \sqrt{z^2 + \rho^2}$
   $\theta = \theta$
   $\phi = \arctan \rho/z$
   To convert from spherical information to cylindrical, we can reverse the previous equations
   $\rho = r \sin \phi$
   $z = r \cos \phi$
   $\theta = \theta$

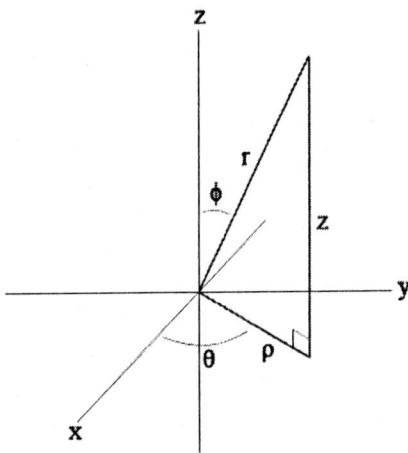

Figure 3.5: Conversion between cylindrical and spherical coordinates.

3. Verify eqn. (3.60).
   Solution:
$$\mathbf{curl F} = \frac{1}{\rho} \begin{vmatrix} \mathbf{e}_\rho & \rho \mathbf{e}_\theta & \mathbf{e}_z \\ \frac{\partial}{\partial \rho} & \frac{\partial}{\partial \theta} & \frac{\partial}{\partial z} \\ \mathbf{F}_\rho & \rho \mathbf{F}_\theta & \mathbf{F}_z \end{vmatrix}$$
$$= \frac{\mathbf{e}_\rho}{\rho} \left(\frac{\partial \mathbf{F}_z}{\partial \theta} - \frac{\partial (\rho \mathbf{F}_\theta)}{\partial z} \right) + \mathbf{e}_\theta \left(\frac{\partial \mathbf{F}_\rho}{\partial z} - \frac{\partial \mathbf{F}_z}{\partial \rho} \right) + \frac{\mathbf{e}_z}{\rho} \left(\frac{\partial (\rho \mathbf{F}_\theta)}{\partial \rho} - \frac{\partial \mathbf{F}_\rho}{\partial \theta} \right)$$
$$= \frac{1}{\rho} \left(\frac{\partial \mathbf{F}_z}{\partial \theta} - \frac{\partial (\mathbf{F}_\theta)}{\partial z} \right) \mathbf{e}_\rho + \left(\frac{\partial \mathbf{F}_\rho}{\partial z} - \frac{\partial \mathbf{F}_z}{\partial \rho} \right) \mathbf{e}_\theta + \frac{1}{\rho} \left(\frac{\partial (\rho \mathbf{F}_\theta)}{\partial \rho} - \frac{\partial \mathbf{F}_\rho}{\partial \theta} \right) \mathbf{e}_z$$

5. Verify eqn. (3.68).
   Solution: Equation (3.68) is $\mathbf{curl F} = \frac{1}{r^2 \sin \phi} \begin{vmatrix} \mathbf{e}_r & r \mathbf{e}_\phi & r \sin \phi \mathbf{e}_\theta \\ \frac{\partial}{\partial r} & \frac{\partial}{\partial \phi} & \frac{\partial}{\partial \theta} \\ F_r & r F_\phi & r \sin \phi F_\theta \end{vmatrix}$. Writing this out in compo-

3.11. ORTHOGONAL CURVILINEAR COORDINATES

nents, we obtain $\frac{1}{r^2 \sin \phi} \left[\mathbf{e}_r \left(r \frac{\partial (\sin \phi F_\theta)}{\partial \phi} - r \frac{\partial F_\phi}{\partial \theta} \right) + r \mathbf{e}_\phi \left(\frac{\partial F_r}{\partial \theta} - \sin \phi \frac{\partial (rF_\theta)}{\partial r} \right) + r \sin \phi \mathbf{e}_\theta \left(\frac{\partial F_\phi}{\partial r} - \frac{\partial F_r}{\partial \phi} \right) \right]$
or $\frac{1}{r \sin \phi} \left[\frac{\partial (\sin \phi F_\theta)}{\partial \phi} - \frac{\partial F_\phi}{\partial \theta} \right] \mathbf{e}_r + \frac{1}{r \sin \phi} \left[\frac{\partial F_r}{\partial \theta} - \sin \phi \frac{\partial (rF_\theta)}{\partial r} \right] \mathbf{e}_\phi + \frac{1}{r} \left[\frac{\partial F_\phi}{\partial r} - \frac{\partial F_r}{\partial \phi} \right] \mathbf{e}_\theta$, which is in accordance with the calculations on page 167 of the text.

7. Show that if f is a function of r only, then $\nabla^2 f(r) = f''(r) + \frac{2}{r} f'(r)$.
   Solution: We know $\nabla^2 f = \nabla \cdot (\nabla f) = \nabla \cdot \left(\frac{\partial f}{\partial r} \mathbf{e}_r \right) = \frac{1}{r^2} \left(\frac{\partial}{\partial r} r^2 \left(\frac{\partial f}{\partial r} \right) \right) = \frac{1}{r^2} \left(2r \frac{\partial f}{\partial r} + r^2 \frac{\partial^2 f}{\partial r^2} \right) = \frac{2}{r} f'(r) + f''(r)$, and so the result is proved.

9. What is the arc length of the curve $r = \sin \phi$, $\theta = \pi/2$, for $0 \leq \phi \leq \pi$?
   Solution: In spherical coordinates $\mathbf{R} = r \cos \theta \sin \phi \mathbf{e}_r + r \sin \theta \sin \phi \mathbf{e}_\phi + r \cos \phi \mathbf{e}_\theta$, and in this case $\mathbf{R} = \cos \theta \sin^2 \phi \mathbf{e}_r + \sin \theta \sin^2 \phi \mathbf{e}_\phi + \sin \phi \cos \phi \mathbf{e}_\theta$. Because $\theta = \frac{\pi}{2}$, $\mathbf{R} = 0 \mathbf{e}_r + \sin^2 \phi \mathbf{e}_\phi + \sin \phi \cos \phi \mathbf{e}_\theta$. Our parameter is ϕ, so $\frac{d\mathbf{R}}{d\phi} = 2 \sin \phi \cos \phi \mathbf{e}_\phi + (\cos^2 \phi - \sin^2 \phi) \mathbf{e}_\theta$, and $\left| \frac{d\mathbf{R}}{d\phi} \right| = \sqrt{4 \sin^2 \phi \cos^2 \phi + \cos^4 \phi + \sin^4 \phi - 2 \cos^2 \phi \sin^2 \phi} = \sqrt{\cos^2 \phi + \sin^2 \phi} = 1$, and $\int_{\phi=0}^{\pi} 1 \, d\phi = \pi$.

11. Compute the gradient, in spherical coordinates, of $f(r, \phi, \theta) = \cos \phi / r^2$.
    Solution: In spherical coordinates, $\nabla f = \frac{\partial f}{\partial r} \mathbf{e}_r + \frac{1}{r} \frac{\partial f}{\partial \phi} \mathbf{e}_\phi + \frac{1}{r \sin \phi} \frac{\partial f}{\partial \theta} \mathbf{e}_\theta = \frac{-2 \cos \phi}{r^3} \mathbf{e}_r - \frac{\sin \phi}{r^3} \mathbf{e}_\phi + 0 \mathbf{e}_\theta$.

13. For what values of n does $\nabla \cdot r^n \mathbf{e}_r = 0$?
    Solution: We know $\nabla \mathbf{F} = \nabla \cdot (r^n \mathbf{e}_r) = \frac{1}{r^2} \frac{\partial}{\partial r} (r^2 r^n) = \frac{1}{r^2} \frac{\partial}{\partial r} r^{2+n} = \frac{1}{r^2} (2+n) r^{1+n}$, so if $n = -2$, $r^{2+n} = r^0 = 1$, a constant, and $\nabla \cdot (r^n \mathbf{e}_r) = 0$.

15. Verify the calculations in example 3.31 [eqs. (2.47) may be helpful.]
    Solution: We must verify the last three lines of display math on page 168

    (a) Because $\rho = (x^2 + y^2)^{\frac{1}{2}}$, $\frac{\partial \rho}{\partial x} = \frac{\frac{1}{2} 2x}{\sqrt{x^2+y^2}} = \frac{x}{\rho} = \cos \theta$.

    (b) Because $\theta = \arctan \frac{y}{x}$, $\frac{\partial \theta}{\partial x} = \frac{1}{1+\left(\frac{y}{x}\right)^2} \frac{\partial}{\partial x} \frac{y}{x} = \frac{-y}{x^2 \left(1+\left(\frac{y}{x}\right)^2\right)} = \frac{-y}{\rho^2} = -\frac{\sin \theta}{\rho}$.

    (c) $\frac{\partial z}{\partial x} = 0$ is obvious.

    (d) $\frac{\partial \rho}{\partial y} = \frac{\frac{1}{2} 2y}{\sqrt{x^2+y^2}} = \frac{y}{\rho} = \sin \theta$.

    (e) $\frac{\partial \theta}{\partial y} = \frac{1}{1+\left(\frac{y}{x}\right)^2} \frac{\partial}{\partial y} \frac{y}{x} = \frac{1}{x \left(1+\left(\frac{y}{x}\right)^2\right)} = \frac{x}{x^2 \left(1+\left(\frac{y}{x}\right)^2\right)} = \frac{x}{\rho^2} = \frac{\cos \theta}{\rho}$.

    (f) $\frac{\partial z}{\partial y} = 0$ is obvious, as is $\frac{\partial \rho}{\partial z} = 0$, $\frac{\partial \theta}{\partial z} = 0$, and $\frac{\partial z}{\partial z} = 1$.

    and the first two equations on page 169 can be obtained by following example 3.28.

3.11 Orthogonal Curvilinear Coordinates

1. Verify that the formulas for the vector operations in cylindrical and spherical coordinates, as computed in section 3.10, are instances of the general formulas derived in this section when the scale factors from example 3.33 are inserted.
   Solution: The scale factors from example 3.33 are h_ρ, h_θ, h_z and h_r, h_ϕ, h_θ for cylindrical and spherical coordinates, respectively. The vector operations are gradient

   $$\nabla f = \frac{1}{h_1} \frac{\partial f}{\partial u_1} \mathbf{e}_1 + \frac{1}{h_2} \frac{\partial f}{\partial u_2} \mathbf{e}_2 + \frac{1}{h_3} \frac{\partial f}{\partial u_3} \mathbf{e}_3,$$

   divergence

   $$\nabla \cdot \mathbf{F} = \frac{1}{h_1 h_2 h_3} \left[\frac{\partial}{\partial u_1} (F_1 h_2 h_3) + \frac{\partial}{\partial u_2} (F_2 h_1 h_3) + \frac{\partial}{\partial u_3} (F_3 h_1 h_2) \right],$$

3.11. ORTHOGONAL CURVILINEAR COORDINATES

curl

$$\nabla \times \mathbf{F} = \frac{1}{h_1 h_2 h_3} \begin{vmatrix} h_1 \mathbf{e}_1 & h_2 \mathbf{e}_2 & h_3 \mathbf{e}_3 \\ \frac{\partial}{\partial u_1} & \frac{\partial}{\partial u_2} & \frac{\partial}{\partial u_3} \\ F_1 h_1 & F_2 h_2 & F_3 h_3 \end{vmatrix},$$

and Laplacian

$$\nabla^2 f = \frac{1}{h_1 h_2 h_3} \left[\frac{\partial}{\partial u_1} \left(\frac{h_2 h_3}{h_1} \frac{\partial f}{\partial u_1} \right) + \frac{\partial}{\partial u_2} \left(\frac{h_1 h_3}{h_2} \frac{\partial f}{\partial u_2} \right) + \frac{\partial}{\partial u_3} \left(\frac{h_1 h_2}{h_3} \frac{\partial f}{\partial u_3} \right) \right].$$

Substituting the scale factors for cylindrical coordinates, we get

$$\nabla f = \frac{1}{1} \frac{\partial f}{\partial \rho} \mathbf{e}_\rho + \frac{1}{\rho} \frac{\partial f}{\partial \theta} \mathbf{e}_\theta + \frac{1}{1} \frac{\partial f}{\partial z} \mathbf{e}_z = \frac{\partial f}{\partial \rho} \mathbf{e}_\rho + \frac{1}{\rho} \frac{\partial f}{\partial \theta} \mathbf{e}_\theta + \frac{\partial f}{\partial z} \mathbf{e}_z,$$

$$\nabla \cdot \mathbf{F} = \frac{1}{\rho} \left[\frac{\partial}{\partial \rho}(F_\rho \rho) + \frac{\partial}{\partial \theta}(F_\theta) + \frac{\partial}{\partial z}(F_z \rho) \right] = \frac{1}{\rho} \frac{\partial}{\partial \rho}(F_\rho \rho) + \frac{1}{\rho} \frac{\partial}{\partial \theta} F_\theta + \frac{\partial}{\partial z} F_z,$$

$$\nabla \times \mathbf{F} = \frac{1}{\rho} \begin{vmatrix} \mathbf{e}_\rho & \rho \mathbf{e}_\theta & \mathbf{e}_z \\ \frac{\partial}{\partial \rho} & \frac{\partial}{\partial \theta} & \frac{\partial}{\partial z} \\ F_\rho & F_\theta \rho & F_z \end{vmatrix},$$

$$\nabla^2 f = \frac{1}{\rho} \left[\frac{\partial}{\partial \rho}\left(\rho \frac{\partial f}{\partial \rho}\right) + \frac{\partial}{\partial \theta}\left(\frac{1}{\rho} \frac{\partial f}{\partial \theta}\right) + \frac{\partial}{\partial z}\left(\rho \frac{\partial f}{\partial z}\right) \right] = \frac{1}{\rho} \frac{\partial}{\partial \rho}\left(\rho \frac{\partial f}{\partial \rho}\right) + \frac{1}{\rho^2} \frac{\partial^2 f}{\partial \theta^2} + \frac{\partial^2 f}{\partial z^2},$$

and for spherical coordinates, we find

$$\nabla f = \frac{1}{1} \frac{\partial f}{\partial r} \mathbf{e}_r + \frac{1}{r} \frac{\partial f}{\partial \phi} \mathbf{e}_\phi + \frac{1}{r \sin \phi} \frac{\partial f}{\partial \theta} \mathbf{e}_\theta = \frac{\partial f}{\partial r} \mathbf{e}_r + \frac{1}{r} \frac{\partial f}{\partial \phi} \mathbf{e}_\phi + \frac{1}{r \sin \phi} \frac{\partial f}{\partial \theta} \mathbf{e}_\theta,$$

$$\nabla \cdot \mathbf{F} = \frac{1}{r^2 \sin \phi} \left[\frac{\partial}{\partial r}(F_r r^2 \sin \phi) + \frac{\partial}{\partial \phi}(F_\phi r \sin \phi) + \frac{\partial}{\partial \theta}(F_\theta r) \right] =$$

$$= \frac{1}{r^2} \frac{\partial}{\partial r}(F_r r^2) + \frac{1}{r \sin \phi} \frac{\partial}{\partial \phi}(F_\phi \sin \phi) + \frac{1}{r \sin \phi} \frac{\partial}{\partial \theta} F_\theta,$$

$$\nabla \times \mathbf{F} = \frac{1}{r^2 \sin \phi} \begin{vmatrix} \mathbf{e}_r & r \mathbf{e}_\phi & r \sin \phi \mathbf{e}_\theta \\ \frac{\partial}{\partial r} & \frac{\partial}{\partial \phi} & \frac{\partial}{\partial \theta} \\ F_r & F_\phi r & F_\theta r \sin \phi \end{vmatrix},$$

$$\nabla^2 f = \frac{1}{r^2 \sin \phi} \left[\frac{\partial}{\partial r}\left(r^2 \sin \phi \frac{\partial f}{\partial r}\right) + \frac{\partial}{\partial \phi}\left(\sin \phi \frac{\partial f}{\partial \phi}\right) + \frac{\partial}{\partial \theta}\left(\frac{1}{\sin \phi} \frac{\partial f}{\partial \theta}\right) \right] =$$

$$= \frac{1}{r^2} \frac{\partial}{\partial r}\left(r^2 \frac{\partial f}{\partial r}\right) + \frac{1}{r^2 \sin \phi} \frac{\partial}{\partial \phi}\left(\sin \phi \frac{\partial f}{\partial \phi}\right) + \frac{1}{r^2 \sin^2 \phi} \frac{\partial^2 f}{\partial \theta^2},$$

3. Explain why curvilinear coordinates defined by functions of the form $u_1 = u_1(z)$, $u_2 = u_2(x)$, $u_3 = u_3(y)$ are automatically orthogonal. What other combinations have this property? How about the following? $u_1 = u_1(\rho)$, $u_2 = u_2(\theta)$, $u_3 = u_3(z)$
Solution: The gradients of the coordinate functions u_i are normal to the isotimic surfaces defined by the u_i and the $\frac{u_i}{|u_i|}$ form the basis vectors $\mathbf{e}_i$ for the coordinate system. The gradients $\nabla u_1(z) = \frac{\partial u_1}{\partial z}\mathbf{k}$, $\nabla u_2(x) = \frac{\partial u_2}{\partial x}\mathbf{i}$, $\nabla u_3(y) = \frac{\partial u_3}{\partial y}\mathbf{j}$ are automatically orthogonal. Any combination for which each u_i is a function of each of the variables x, y, z separately will also have this property. For the suggested set $u_1 = u_1(\rho)$, $u_2 = u_2(\theta)$, $u_3 = u_3(z)$ which is a new coordinate system expressed in cylindrical coordinates, we must calculate $\nabla u_1(\rho) = \frac{\partial u_1}{\partial \rho}\mathbf{e}_\rho$, $\nabla u_2(\theta) = \frac{1}{\rho}\frac{\partial u_2}{\partial \theta}\mathbf{e}_\theta$, $\nabla u_3(z) = \frac{\partial u_3}{\partial z}\mathbf{e}_z$, which are also orthogonal in a cylindrical coordinate system.

3.11. ORTHOGONAL CURVILINEAR COORDINATES

5. Compute $\nabla^2 g$ if $g = u_1^3 + u_2^3 + u_3^3$ in the coordinate system in the coordinate system in equation 3.78.
   Solution: The coordinate system of equation 3.78 is $x = u_1 - u_2$, $y = u_1 + u_2$, $z = u_3^2$. Solve for the u_i to get $u_1 = \frac{x+y}{2}$, $u_2 = \frac{y-x}{2}$, $u_3 = z^{1/2}$. The scale factors are then $h_1 = \sqrt{2}$, $h_2 = \sqrt{2}$, $h_3 = 2u_3$. The Laplacian of g is then

$$\nabla^2 g = \frac{1}{4u_3}\left[\frac{\partial}{\partial u_1}\left(2u_3 \frac{\partial g}{\partial u_1}\right) + \frac{\partial}{\partial u_2}\left(2u_3 \frac{\partial g}{\partial u_2}\right) + \frac{\partial}{\partial u_3}\left(\frac{1}{u_3}\frac{\partial g}{\partial u_3}\right)\right] =$$

$$= \frac{1}{4u_3}\left[\frac{\partial}{\partial u_1}\left(2u_3 \; 3u_1^2\right) + \frac{\partial}{\partial u_2}\left(2u_3 \; 3u_2^2\right) + \frac{\partial}{\partial u_3}\left(\frac{1}{u_3}\; 3u_3^2\right)\right] = (12u_1 u_3 + 12 u_2 u_3 + 3)/4u_3$$

7. Let $u_1 = x + y$, $u_2 = x - 2y$, $u_3 = 2z$.

   (a) Solve for x, y, and z in terms of u_1, u_2, and u_3.
   (b) Attempt to determine the scale factors h_1, h_2, and h_3.
   (c) What is "wrong"?

   Solution:

   (a) $x = \frac{2u_1 + u_2}{3}$, $y = \frac{u_1 - u_2}{3}$, $z = \frac{1}{2}u_3$.

   (b) Recall that there are two ways to find scale factors, $h_i = \left|\frac{\partial \mathbf{R}}{\partial u_i}\right|$ when we have $\mathbf{R} = x(u_1, u_2, u_3)\mathbf{i} + y(u_1, u_2, u_3)\mathbf{j} + z(u_1, u_2, u_3)\mathbf{k}$ and $h_i = \frac{1}{|\nabla u_i|}$ when we have $u_i = u_i(x, y, z)$. In this case we have both available, so we calculate both ways (from the question statement first, then from the solution in (a)):

   (a) $h_1 = \frac{\sqrt{2}}{2}$, $h_2 = \frac{\sqrt{5}}{5}$, $h_3 = \frac{\sqrt{2}}{2}$.
   (b) $h_1 = \frac{\sqrt{5}}{3}$, $h_2 = \frac{\sqrt{2}}{3}$, $h_3 = \frac{1}{2}$.

   (c) What is "wrong"? The system is not orthogonal, so the outward normals to the isotimic surfaces do not point in the same directions as the tangents to the coordinate curves. Therefore we cannot expect the calculations for the scale factors to agree.

9. Consider the transformation $x = u_3$, $y = e^{u_2} \cos u_1$, $z = e^{u_2} \sin u_1$.

   (a) Show that (u_1, u_2, u_3) constitute orthogonal curvilinear coordinates.
   (b) Compute the scale factors.
   (c) Find $\nabla^2 g$ if $g = u_1^2 + u_2^2 + u_3^2$
   (d) Find the divergence and curl of the vector field $\mathbf{F} = -e^{u_2}\mathbf{e}_3 + u_3\mathbf{e}_1$.

   Solution:

   (a) Recall that if we have $u_i = u_i(x, y, z)$, we can tell (u_1, u_2, u_3) are orthogonal if $\nabla u_i \cdot \nabla u_j = 0$ for $i \neq j$. On the other hand, if we have $x = x(u_1, u_2, u_3), y = y(u_1, u_2, u_3), z = z(u_1, u_2, u_3)$ we test for orthogonality by computing $\frac{\partial \mathbf{R}}{\partial u_i} \cdot \frac{\partial \mathbf{R}}{\partial u_j} = 0$ for $i \neq j$. We have the second situation, so we find $\frac{\partial \mathbf{R}}{\partial u_i}$.

   (a) $\frac{\partial \mathbf{R}}{\partial u_1} = 0\mathbf{e}_1 - e^{u_2}\sin u_1 \mathbf{e}_2 + e^{u_2}\cos u_1 \mathbf{e}_3$
   (b) $\frac{\partial \mathbf{R}}{\partial u_2} = 0\mathbf{e}_1 + e^{u_2}\cos u_1 \mathbf{e}_2 + e^{u_2}\sin u_1 \mathbf{e}_3$
   (c) $\frac{\partial \mathbf{R}}{\partial u_3} = 1\mathbf{e}_1 + 0\mathbf{e}_2 + 0\mathbf{e}_3$

   It is easy to see that this set is orthogonal.

   (b) The scale factors $h_i = \left|\frac{\partial \mathbf{R}}{\partial u_i}\right|$ are found using the tangent vectors to the coordinate curves we found in the first part of the problem. We find $h_1 = e^{u_2}$, $h_2 = e^{u_2}$, $h_3 = 1$.

   (c) The laplacian is

$$\nabla^2 g = e^{-2u_2}\left[\frac{\partial}{\partial u_1}\left(\frac{\partial g}{\partial u_1}\right) + \frac{\partial}{\partial u_2}\left(\frac{\partial g}{\partial u_2}\right) + \frac{\partial}{\partial u_3}\left(e^{2u_2}\frac{\partial g}{\partial u_3}\right)\right] =$$

$$\nabla^2 g = e^{-2u_2}\left[\frac{\partial}{\partial u_1}(2u_1) + \frac{\partial}{\partial u_2}(2u_2) + \frac{\partial}{\partial u_3}\left(e^{2u_2}\; 2u_3\right)\right] = 2 + 4e^{-2u_2}$$

(d) The divergence is $\nabla \cdot \mathbf{F} = \frac{1}{h_1 h_2 h_3}\left[\frac{\partial}{\partial u_1}(F_1 h_2 h_3) + \frac{\partial}{\partial u_2}(F_2 h_1 h_3) + \frac{\partial}{\partial u_3}(F_3 h_2 h_1)\right]$, the curl is

$$\nabla \times \mathbf{F} = \frac{1}{h_1 h_2 h_3}\begin{vmatrix} h_1 \mathbf{e}_1 & h_2 \mathbf{e}_2 & h_3 \mathbf{e}_3 \\ \frac{\partial}{\partial u_1} & \frac{\partial}{\partial u_2} & \frac{\partial}{\partial u_3} \\ F_1 h_1 & F_2 h_2 & F_3 h_3 \end{vmatrix},$$

and we found the scale factors to be $h_1 = e^{u_2}$, $h_2 = e^{u_2}$, $h_3 = 1$ so $\nabla \cdot \mathbf{F} = e^{-2u_2}\left[\frac{\partial}{\partial u_1}(u_3 e^{u_2}) + \frac{\partial}{\partial u_2}(0 e^{u_2}) + \frac{\partial}{\partial u_3}(-e^{u_2} e^{2u_2})\right] = 0$ and

the curl is $\nabla \times \mathbf{F} = e^{-2u_2}\begin{vmatrix} e^{u_2}\mathbf{e}_1 & e^{u_2}\mathbf{e}_2 & \mathbf{e}_3 \\ \frac{\partial}{\partial u_1} & \frac{\partial}{\partial u_2} & \frac{\partial}{\partial u_3} \\ u_3 e^{u_2} & 0 & -e^{u_2} \end{vmatrix} e^{-2u_2}(-e^{2u_2}\mathbf{e}_1 + e^{2u_2}\mathbf{e}_2 - u_3 e^{u_2}\mathbf{e}_3) =$

$= -\mathbf{e}_1 + \mathbf{e}_2 - u_3 e^{-u_2}\mathbf{e}_3.$

11. Suppose that u, v, w are orthogonal curvilinear coordinates for which $ds^2 = v^2 du^2 + u^2 dv^2 + dw^2$.

    (a) Calculate the divergence of $\mathbf{u}$, where $\mathbf{u}$ is the unit vector tangent to a u curve.

    (b) Determine the laplacian of the function $f = uvw$.

    Solution:

    (a) Arc length in general orthogonal curvilinear coordinates is given by $ds = \left(h_1^2 du_1^2 + h_2^2 du_2^2 + h_3^2 du_3^2\right)^{1/2}$, so we know the scale factors for this coordinate system are $h_1 = v$, $h_2 = u$ and $h_3 = 1$. The scale factors are given by $h_i = \left|\frac{\partial \mathbf{R}}{\partial u_i}\right|$, and a tangent $\mathbf{F}$ to a u curve is $\mathbf{F} = \frac{\partial \mathbf{R}}{\partial u_1}\mathbf{e}_1 + \frac{\partial \mathbf{R}}{\partial u_2}\mathbf{e}_2 + \frac{\partial \mathbf{R}}{\partial u_3}\mathbf{e}_3 = \frac{\partial \mathbf{R}}{\partial u_1}\mathbf{e}_u + \frac{\partial \mathbf{R}}{\partial u_2}\mathbf{e}_v + \frac{\partial \mathbf{R}}{\partial u_3}\mathbf{e}_w = \frac{\partial \mathbf{R}}{\partial u_1}\mathbf{e}_u$ because the components of of the tangent vector to a u curve in the v and w directions are 0. The divergence of $\mathbf{F}$ is
    $\nabla \cdot \mathbf{F} = \frac{1}{h_1 h_2 h_3}\left[\frac{\partial}{\partial u_1}(F_1 h_2 h_3) + \frac{\partial}{\partial u_2}(F_2 h_1 h_3) + \frac{\partial}{\partial u_3}(F_3 h_2 h_1)\right]$
    $= \frac{1}{uv}\left[\frac{\partial}{\partial u_1}(F_1 h_2 h_3) + \frac{\partial}{\partial u_2}(F_2 h_1 h_3) + \frac{\partial}{\partial u_3}(F_3 h_2 h_1)\right] = \frac{1}{uv}\left[\frac{\partial}{\partial u}(1u)\right] = \frac{1}{uv}$, where we have used the fact that the component in the direction of the u curve of a unit tangent vector must be 1.

    (b) The laplacian in general orthogonal curvilinear coordinates is

    $$\nabla^2 \cdot f = \frac{1}{h_1 h_2 h_3}\left[\frac{\partial}{\partial u_1}\left(\frac{h_2 h_3}{h_1}\frac{\partial f}{\partial u_1}\right) + \frac{\partial}{\partial u_2}\left(\frac{h_1 h_3}{h_2}\frac{\partial f}{\partial u_2}\right) + \frac{\partial}{\partial u_3}\left(\frac{h_2 h_1}{h_3}\frac{\partial f}{\partial u_3}\right)\right] =$$

    $$= \frac{1}{uv}\left[\frac{\partial}{\partial u}\left(\frac{u}{v}vw\right) + \frac{\partial}{\partial v}\left(\frac{v}{u}uw\right) + \frac{\partial}{\partial w}\left(u^2 v^2\right)\right] = 2w/uv.$$

13. What is the element of volume in parabolic cylindrical coordinates?
    Solution: The parabolic cylindrical coordinate system (see problem 12 in this section) is defined by $x = 1/2(u^2 - v^2)$, $y = uv$ and $z = z$. The scale factors are $h_u = h_v = \sqrt{u^2 + v^2}$ and $h_z = 1$, so the volume element is $(u^2 + v^2) du\, dv\, dz$.

3.11. ORTHOGONAL CURVILINEAR COORDINATES

Chapter 4

Line, Surface, and Volume Integrals

4.1 Line Integrals

1. Find $\int \mathbf{F} \cdot d\mathbf{R}$, where $\mathbf{F} = x^2\mathbf{i} + \mathbf{j} + yz\mathbf{k}$ along $C : x = t,\ y = 2t^2,\ z = 3t,\ 0 \leq t \leq 1$.
   Solution: The path is given by $t\mathbf{i} + 2t^2\mathbf{j} + 3t\mathbf{k}$, and along this path the vector field is $\mathbf{F} = t^2\mathbf{i} + \mathbf{j} + 6t^3\mathbf{k}$. Then $\mathbf{F} \cdot \frac{d\mathbf{R}}{dt} = (t^2\mathbf{i} + \mathbf{j} + 6t^3\mathbf{k}) \cdot (\mathbf{i} + 4t\mathbf{j} + 3\mathbf{k}) = t^2 + 4t + 18t^3$. Integrating this gives $\int_{t=0}^{t=1} t^2 + 4t + 18t^3\, dt = \frac{41}{6}$.

3. Find $\int \mathbf{F} \cdot d\mathbf{R}$ from $(1,0,0)$ to $(1,0,4)$ if $\mathbf{F} = x\mathbf{i} - y\mathbf{j} + z\mathbf{k}$.

   (a) along the line segment joining $(1,0,0)$ and $(1,0,4)$.

   (b) along the helix $x = \cos 2\pi t,\ y = \sin 2\pi t,\ z = 4t$.

   Solution:

   (a) Along the line segment joining $(1,0,0)$ and $(1,0,4)$, $\mathbf{R} = \mathbf{i} + 4t\mathbf{k}$ and along this path $\mathbf{F} = \mathbf{i} + 4t\mathbf{k}$. Thus we are integrating $\int_{t=0}^{t=1} 16t\, dt = 8$.

   (b) Along the helix $\mathbf{R} = \cos 2\pi t\, \mathbf{i} + \sin 2\pi t\, \mathbf{j} + 4t\mathbf{k}$, $\mathbf{F} = \cos 2\pi t\, \mathbf{i} - \sin 2\pi t\, \mathbf{j} + 4t\mathbf{k}$, so $\int_{t=0}^{t=1} \mathbf{F} \cdot \frac{d\mathbf{R}}{dt}\, dt = \int_{t=0}^{t=1} -4\pi \cos 2\pi t \sin 2\pi t + 16t\, dt = 8$.

5. Find the line integral $\int \mathbf{F} \cdot d\mathbf{R}$ along the line segment from $(1,0,2)$ to $(3,4,1)$, where $\mathbf{F} = 2xy\mathbf{i} + (x^2 + z)\mathbf{j} + y\mathbf{k}$.
   Solution: The position vector $\mathbf{R} = (1 + 2t)\mathbf{i} + (4t)\mathbf{j} + (1 - t)\mathbf{k}$ traces out the line segment as t goes from 0 to 1, and along that segment, $\mathbf{F} = (8t + 16t^2)\mathbf{i} + (2 + 3t + 4t^2)\mathbf{j} + 4t\mathbf{k}$. We integrate $\int_{t=0}^{t=1} \mathbf{F} \cdot \frac{d\mathbf{R}}{dt}\, dt = \int_{t=0}^{t=1} (2\mathbf{i} + 4\mathbf{j} - \mathbf{k}) \cdot ((8t + 16t^2)\mathbf{i} + (3 + 3t + 4t^2)\mathbf{j} + (4t)\mathbf{k})\, dt = 40$.

7. Let
$$\mathbf{F} = \frac{y}{x^2 + y^2}\mathbf{i} - \frac{x}{x^2 + y^2}\mathbf{j}.$$

   Find the line integral of the tangential component of $\mathbf{F}$ from $(-1, 0)$ to $(1, 0)$

   (a) along the semicircle $y = \sqrt{1 - x^2}$.

   (b) along the dotted polygonal path shown in figure 4.4.

   Solution:

4.1. LINE INTEGRALS

(a) We can think of x as our parameter here, so the position vector traces out the semi-circle $\mathbf{R} = x\mathbf{i} + \sqrt{1-x^2}\mathbf{j}$ as x varies from -1 to 1, and the tangent $\frac{d\mathbf{R}}{dx} = \mathbf{i} - \frac{x}{\sqrt{1-x^2}}\mathbf{j}$ so $\mathbf{F} \cdot \frac{d\mathbf{R}}{dx} = \frac{1}{\sqrt{1-x^2}}$ and $\int \mathbf{F} \cdot \frac{d\mathbf{R}}{dx} dx = \int_{-1}^{1} \frac{dx}{\sqrt{1-x^2}} = \pi$.

(b) For the second part, it is more convenient to use a path with a parameter t to accommodate the vertical segment in the path. Computing the projection of $\mathbf{F}$ along the paths $\mathbf{R}_1 = t\mathbf{i} + (t+1)\mathbf{j}$, $\mathbf{R}_2 = t\mathbf{i} + \mathbf{j}$ and $\mathbf{R}_3 = \mathbf{i} + (1-t)\mathbf{j}$ as t varies from 0 to 1 along *each* of the three segments. Thus we compute $\int_{t=0}^{t=1} \mathbf{F} \cdot \frac{d\mathbf{R}}{dt} dt = \int_{t=0}^{t=1} \mathbf{F}_1 \cdot \frac{d\mathbf{R}_1}{dt} + \mathbf{F}_2 \cdot \frac{d\mathbf{R}_2}{dt} + \mathbf{F}_3 \cdot \frac{d\mathbf{R}_3}{dt} dt = \pi$, where the $\mathbf{F}_i$, $i = 1, 2, 3$ are the vector fields along each segment.

9. Evaluate the following line integrals over the straight line segment C joining the point $(2, 1, 4)$ to the point $(3, 3, 4)$

    (a) $\int_C [3xy\, dx + 3\, dy + yz\, dz]$

    (b) $\int_C e^{xyz}(yz\, dx + xz\, dy + xy\, dz)$

    Solution:

    (a) Make a linear interpolation between the points to get the position vector that traces out the line segment: $\mathbf{R} = (1-t)(2\mathbf{i}+\mathbf{j}+4\mathbf{k}) + t(3\mathbf{i}+3\mathbf{j}+4\mathbf{k}) = (2+t)\mathbf{i}+(1+2t)\mathbf{j}+4\mathbf{k}$ and its derivative is $\frac{d\mathbf{R}}{dt} = \mathbf{i}+2\mathbf{j}$. The integral written this way $\int_C [f_1\, dx + f_2\, dy + f_3\, dz]$ can also be interpreted as $\int \frac{d\mathbf{R}}{dt} \cdot \mathbf{F} dt = \int \frac{dx}{dt}f_1 + \frac{dy}{dt}f_2 + \frac{dz}{dt}f_3 dt$, so in this case, we have $\int_0^1 3(2+t)(1+2t)+6 dt = 43/2$

    (b) Using the same path, $\int_C e^{xyz}(yz\, dx + xz\, dy xy\, dz)$ becomes $\int_C e^{(2+t)(1+2t)4}((1+2t)4 + (2+t)(4)2) dt$. Note that the part multiplying the exponential is the derivative of the exponent, so the integral is just $e^{8+20t+8t^2} |_0^1 = e^{36} - e^8$.

11. Evaluate $\oint_C [(\sin x + y^2)\, dx + (x - e^{-y})\, dy]$, where C is the boundary of the semi-circular region $x^2 + y^2 \leq 4$, $y \geq 0$.
    Solution: The solution is probably easiest in polar coordinates, although cartesian coordinates can be used. Along the path, the parameter is θ, and $x = 2\cos\theta$, $y = 2\sin\theta$, $dx = -2\sin\theta d\theta$, $dy = 2\cos\theta d\theta$. Then we must integrate

    $$\int_{\theta=-\pi}^{\theta=0} \left[(\sin 2\cos\theta + 4\sin^2\theta)(-2\sin\theta d\theta) + (2\cos\theta - e^{-2\sin\theta})(2\cos\theta d\theta) \right].$$

    We must make a u substitution for the arguments of the first sine function and the exponential, so breaking up the integral gives us $\int_{(\theta=-\pi)}^{(\theta=0)} \sin u\, du = 0$, $\int_{\theta=-\pi}^{\theta=0} -8\sin^3\theta d\theta = 32/3$, $\int_{\theta=-\pi}^{\theta=0} 4\cos^2\theta d\theta = 2\pi$, $\int_{(\theta=-\pi)}^{(\theta=0)} e^{-u} du = 0$, where we have put the limits for the integration in changed variables in parentheses to indicate that they are not current. Thus the integrals sum to $32/3 + 2\pi$.

13. Find $\int \mathbf{R} \cdot d\mathbf{R}$ from $(1, 2, 2)$ to $(3, 6, 6)$, along the line segment joining these points,

    (a) in the manner described in the text.

    (b) by observing that $\mathbf{R} \cdot d\mathbf{R} = s\, ds$, where $s = (x^2 + y^2 + z^2)^{1/2}$ is the distance from the origin, and computing $\int_3^9 s\, ds$.

    Solution:

    (a) Here $\mathbf{R} = (1+2t)\mathbf{i}+(2+4t)\mathbf{j}+(2+4t)\mathbf{k}$, $\frac{d\mathbf{R}}{dt} = 2\mathbf{i}+4\mathbf{j}+4\mathbf{k}$ and $\mathbf{R} \cdot \frac{d\mathbf{R}}{dt} = 18+36t$, so the integral is $\int_0^1 18 + 36t dt = 36$. Note that there are many ways to get a parameterized line segment. We prefer the simple method of linear interpolation: $\mathbf{R} = (1-t)\mathbf{v}_1 + t\mathbf{v}_2$. Alternatively, you can compute the difference vector $2\mathbf{i}+4\mathbf{j}+4\mathbf{k}$ and add t times this to $\mathbf{i}+2\mathbf{j}+2\mathbf{k}$ to get the same thing.

    (b) Because the position vector points in the same direction as its derivative, we can compute the integral as if it were distance along a straight line (it is!). The limits of integration will be the distance of the two points from the origin, $\sqrt{1^2+2^2+2^2} = 3$ and $\sqrt{3^2+6^2+6^2} = 9$, so we compute $\int_3^9 s\, ds = \frac{1}{2}s^2|_3^9 = \frac{1}{2}(81-9) = 36$.

54

15. For example 4.3 (refer to figure 4.4), determine $\mathbf{T}$,

   (a) along path (1), in the direction shown in terms of $\mathbf{i}$ and $\mathbf{j}$.

   (b) along path (2) in the direction shown.

   (c) along path (3) in the direction shown.

   Solution:

   (a) Along path (1), $\mathbf{R} = x\mathbf{i} + (x+1)\mathbf{j}$ so $\mathbf{T} = \frac{\frac{d\mathbf{R}}{dt}}{|\frac{d\mathbf{R}}{dt}|} = \frac{\sqrt{2}}{2}(\mathbf{i}+\mathbf{j})$.

   (b) Along path (2), $\mathbf{R} = x\mathbf{i} + \mathbf{j}$ so $\mathbf{T} = \frac{\frac{d\mathbf{R}}{dt}}{|\frac{d\mathbf{R}}{dt}|} = \mathbf{i}$.

   (c) Along path (3), $\mathbf{R} = -y\mathbf{j}$ so $\mathbf{T} = \frac{\frac{d\mathbf{R}}{dt}}{|\frac{d\mathbf{R}}{dt}|} = -\mathbf{j}$.

17. Show that $d\mathbf{R} = dx\mathbf{i} + dy\mathbf{j}$ is the same as $\mathbf{T}\,ds$ in each of the three special cases referred to in exercises 15 and 16. (This illustrates the general rule that, in practice, it is easier to find $d\mathbf{R}$ directly than to find $\mathbf{T}$ and ds separately and multiply.)
   Solution: In this case of a two dimensional space and using one of the coordinates as a parameter, we have the equivalence $\frac{d\mathbf{R}}{dx} = \frac{d}{dx}R_1\mathbf{i} + \frac{d}{dx}R_2\mathbf{j} \Rightarrow d\mathbf{R} = dR_1\mathbf{i} + dR_2\mathbf{j}$. Therefore we have

   (a) Along path (1) $d\mathbf{R} = dR_1\mathbf{i} + dR_2\mathbf{j} = dx\mathbf{i} + dx\mathbf{j} = \frac{\sqrt{2}}{2}(\mathbf{i}+\mathbf{j})\sqrt{2} = \mathbf{T}\,ds$, because the distance along the line segment changes at $\sqrt{2}$ times the change with the parameter x.

   (b) Along path (2) $d\mathbf{R} = dR_1\mathbf{i} + dR_2\mathbf{j} = dx\mathbf{i} = \mathbf{T}\,ds$, because the distance along the line segment changes at the same rate as the parameter x.

   (c) Along path (3) $d\mathbf{R} = dR_1\mathbf{i} + dR_2\mathbf{j} = dy\mathbf{j} = \mathbf{T}\,ds$, because the distance along the line segment changes at the same rate as the parameter y.

19. Let $\mathbf{F}(x,y) = (x^2 + y^2)(\mathbf{i}+\mathbf{j})$, and let C be a directed straight line segment of unit length, with one endpoint at the origin $(0,0)$. Find the direction of C such that the line integral $I = \int_C \mathbf{F} \cdot d\mathbf{R}$ is

   (a) a maximum (give the direction of C and the value of I).

   (b) a minimum (give the direction of C and the value of I).

   (c) zero (give the direction of C).

   Solution: The hint here is that we must vary the direction of the line long which we plan to integrate. The most convenient way to do this is to write everything in polar coordinates and use the angle θ as a parameter to change the direction of the path of integration. So let $x = r\cos\theta$, $y = r\sin\theta$. To integrate along the line of unit length, we let r vary from 0 to 1, and the integration will be with respect to dr, so $dx = dr\cos\theta$ and $dy = dr\sin\theta$. Then $\mathbf{R} = r\cos\theta\mathbf{i} + r\sin\theta\mathbf{j}$, $d\mathbf{R} = dr\cos\theta\mathbf{i} + dr\sin\theta\mathbf{j}$, and $\mathbf{F} = r^2\mathbf{i} + r^2\mathbf{j}$. We integrate $\int_{r=0}^{r=1} \mathbf{F}\cdot d\mathbf{R} = \int_{r=0}^{r=1} r^2 dr(\cos\theta + \sin\theta) = \frac{1}{3}(\cos\theta + \sin\theta)$. The maximum $\frac{1}{3}\sqrt{2}$ will occur for $\theta = \frac{\pi}{4}$, the minimum $-\frac{1}{3}\sqrt{2}$ at $\theta = \frac{5\pi}{4}$ and there will be two zeroes at $\theta = \frac{3\pi}{4}$ and $\theta = \frac{7\pi}{4}$.

4.2 Domains: Simply Connected Domains

In each of the following cases, a region D is defined. Tell whether the region is a domain. If it is a domain, determine whether or not it is simply connected. If it is not a domain, explain why not.

1. The region of definition of a magnetic field due to a steady current flowing along the z axis [i.e., the region consisting of all points (x,y,z) such that $x^2 + y^2 \geq 0$].
   Solution: This region is both open (it is an open set $\mathbb{R}^3$ less a closed set, the z axis) and connected, so it is a domain. It is not simply connected because a loop around the z axis cannot be shrunk to a point.

4.3. CONSERVATIVE FIELDS: THE POTENTIAL FUNCTION

3. The region consisting of all points above the xy plane [i.e, all points (x,y,z) such that $z > 0$].
   Solution: This region is open because it does not include the xy plane, its only possible boundary (it is unbounded in the positive z direction and positive and negative x and y directions). It is connected because each point can be joined to any other point by a smooth curve. Because it is open and connected, it is a domain. It is also simply connected because each loop entirely above the xy plane can be shrunk to a point: there are no obstructions. In three dimensions, there are no finite obstructions, because any loop could be moved around the obstruction then shrunk. However, if we subtracted, say, the z axis from this space, then a loop wrapped around the z axis could not be shrunk to a point.

5. The region D consisting of all points (x,y,z) such that $x^2 + y^2 + z^2 > 4$.
   Solution: This is a domain because it is open (it is an open set $\mathbb{R}^3$ less a closed set, the closed sphere of radius 2) and connected. It is simply connected – any loop around the sphere can be slipped off and then shrunk to a point.

7. The region D consisting of all points (x,y,z) for which $1 < x < 2$ [i.e., all points between the planes $x = 1$ and $x = 2$.
   Solution: This region is open because it does not include the planes $x = 1$ and $x = 2$, and of course there is no boundary at infinity in the positive or negative y and z directions. It is connected because each point between the planes $x = 1$ and $x = 2$ can be joined to any other point in the region by a smooth curve. Because it is open and connected, it is a domain. It is also simply connected because each loop lying between the planes $x = 1$ and $x = 2$ can be shrunk to a point: there are no obstructions.

9. The region in the plane between two concentric circles.
   Solution: This is a domain, and it is open (it does not contain the circles that form its boundary) and it is connected, but it is not simply connected because a loop around the inner circle cannot be shrunk to a point.

4.3 Conservative Fields: The Potential Function

1. Show that if $\oint_C \mathbf{F} \cdot d\mathbf{R} = 0$ for every regular closed curve C, then for any two points P and Q, $\int_Q^P \mathbf{F} \cdot d\mathbf{R}$ is independent of path. (*Hint:* Let C_1 and C_2 be two paths extending from P to Q, and construct a closed curve out of these.)
   Solution: Let C_1 and C_2 be two paths joining P and Q, and let C be the path going from P to Q along C_1 then back to P along C_2. Break up the integral $\oint_C \mathbf{F} \cdot d\mathbf{R} = 0$ into $\oint_C \mathbf{F} \cdot d\mathbf{R} = \int_P^Q C_1 \mathbf{F} \cdot d\mathbf{R} + \int_Q^P C_2 \mathbf{F} \cdot d\mathbf{R} = 0$, so $\int_P^Q C_1 \mathbf{F} \cdot d\mathbf{R} = -\int_Q^P C_2 \mathbf{F} \cdot d\mathbf{R}$. Now fix the curve C_1 so that $\int_P^Q C_1 \mathbf{F} \cdot d\mathbf{R} = K$ a constant. Then $\int_Q^P C_2 \mathbf{F} \cdot d\mathbf{R} = K$, and because C_2 was arbitrary, the integral of $\mathbf{F}$ is independent of path.

3. Using methods similar to that of example 4.5, show that the fields of exercise 2 are not conservative.
   Solution: For the two dimensional problems, two separate paths between $(0,0)$ and $(1,1)$ are $\mathbf{R}_1 = t\mathbf{i} + t\mathbf{j}$ and $\mathbf{R}_2 = t\mathbf{i} + t^2\mathbf{j}$, with derivatives $\frac{d\mathbf{R}_1}{dt} = \mathbf{i} + \mathbf{j}$ and $\frac{d\mathbf{R}_2}{dt} = \mathbf{i} + 2t\mathbf{j}$. For the three dimensional problems paths $\mathbf{R}_3 = t\mathbf{i} + t\mathbf{j} + t\mathbf{k}$ and $\mathbf{R}_4 = t\mathbf{i} + t^2\mathbf{j} + t^3\mathbf{k}$ join $(0,0,0)$ and $(1,1,1)$, with tangents $\frac{d\mathbf{R}_3}{dt} = \mathbf{i} + \mathbf{j} + \mathbf{k}$ and $\frac{d\mathbf{R}_4}{dt} = \mathbf{i} + 2t\mathbf{j} + 3t^2\mathbf{k}$. For the final path we go clockwise and counterclockwise around the unit circle around the origin in the plane from $(1,0)$ to $(-1,0)$ along the paths $\mathbf{R}_5 = \cos t\mathbf{i} + \sin t\mathbf{j}$ and $\mathbf{R}_6 = \cos t\mathbf{i} - \sin t\mathbf{j}$.

   (a) Along path one, $\mathbf{F} = -y\mathbf{i} + x\mathbf{j}$ is $\mathbf{F} = -t\mathbf{i} + t\mathbf{j}$, and the integral $\int_0^1 \mathbf{F} \cdot \frac{d\mathbf{R}}{dt} dt = \int_0^1 0 dt = 0$, while along path two, $\mathbf{F} = -t^2\mathbf{i} + t\mathbf{j}$ and the integral is $\int_0^1 \mathbf{F} \cdot \frac{d\mathbf{R}}{dt} dt = \int_0^1 t^2 dt = \frac{1}{3}t^3|_0^1 = \frac{1}{3}$, so the field is not conservative.

   (b) Along path one, $\mathbf{F} = y\mathbf{i} + y(x-1)\mathbf{j}$ is $\mathbf{F} = t\mathbf{i} + t(t-1)\mathbf{j}$, and the integral $\int_0^1 \mathbf{F} \cdot \frac{d\mathbf{R}}{dt} dt = \int_0^1 t^2 dt = 1/3 t^3 |_0^1 = 1/3$, while along path two, $\mathbf{F} = -t^2\mathbf{i} + t^2(t-1)\mathbf{j}$ and the integral is $\int_0^1 \mathbf{F} \cdot \frac{d\mathbf{R}}{dt} dt = \int_0^1 2t^4 - 2t^3 + t^2 dt = \frac{2}{5}t^5 - \frac{1}{2}t^4 + \frac{1}{3}t^3|_0^1 = \frac{7}{30}$, so the field is not conservative.

(c) Along path three, $\mathbf{F} = y\mathbf{i} + x\mathbf{j} + x^2\mathbf{k}$ is $\mathbf{F} = t\mathbf{i} + t\mathbf{j} + t^2\mathbf{k}$, and the integral $\int_0^1 \mathbf{F} \cdot \frac{d\mathbf{R}}{dt} dt = \int_0^1 2t + t^2 dt = t + \frac{1}{3}t^3|_0^1 = 4/3$, while along path four, $\mathbf{F} = t^2\mathbf{i} + t\mathbf{j} + t^2\mathbf{k}$ and the integral is $\int_0^1 \mathbf{F} \cdot \frac{d\mathbf{R}}{dt} dt = \int_0^1 3t^2 + 3t^4 dt = t^3 + \frac{3}{5}t^5|_0^1 = \frac{8}{5}$, so the field is not conservative.

(d) Along path three, $\mathbf{F} = z\mathbf{i} + z\mathbf{j} + (y-1)\mathbf{k}$ is $\mathbf{F} = t\mathbf{i} + t\mathbf{j} + (t-1)\mathbf{k}$, and the integral $\int_0^1 \mathbf{F} \cdot \frac{d\mathbf{R}}{dt} dt = \int_0^1 3t - 1 dt = \frac{3}{2}t^2 - t = \frac{1}{2}$, while along path four, $\mathbf{F} = t^3\mathbf{i} + t^3\mathbf{j} + (t^2-1)\mathbf{k}$ and the integral is $\int_0^1 \mathbf{F} \cdot \frac{d\mathbf{R}}{dt} dt = \int_0^1 5t^4 + t^3 - 3t^2 dt = t^5 + \frac{1}{4}t^4 - t^3|_0^1 = \frac{1}{4}$, so the field is not conservative.

(e) Counterclockwise along path five, $\mathbf{F} = \frac{x}{x^2+y^2}\mathbf{i} + \frac{x}{x^2+y^2}\mathbf{j}$ is $\mathbf{F} = \cos t\mathbf{i} + \cos t\mathbf{j}$, and the integral $\int_0^\pi \mathbf{F} \cdot \frac{d\mathbf{R}}{dt} dt = \int_0^\pi \cos^2 t - \cos t \sin t \, dt = \frac{\pi}{2}$, while clockwise along path six the integral is $\int_0^\pi \mathbf{F} \cdot \frac{d\mathbf{R}}{dt} dt = \int_0^\pi -\cos^2 t - \cos t \sin t \, dt = -\frac{\pi}{2}$, so the field is not conservative.

5. If you worked correctly, you obtained a nonzero answer to exercise 4. Yet it appears that $\mathbf{F} = \text{grad}\phi$, where $\phi = \tan^{-1}(y/x)$, and this would contradict theorem 4.2. Investigate this mystery.
   Solution: Theorem 4.2 states *A vector field $\mathbf{F}$ continuous in a domain D is conservative if and only if around every regular closed curve in D the line integral of $\mathbf{F}$ is zero*. The function $\phi = \tan^{-1}\frac{y}{x}$ is not defined at the origin (nor is the gradient), so the origin is not in the domain D, and any path around the origin is not guaranteed to give a zero line integral. In fact, ϕ is just the angle of the position vector referenced to the x axis, so the path integral of the gradient of ϕ around the origin *cannot* be zero for *any* closed path. However, it is easy to see that if our path does not go around the origin, the integral must be zero, because the total angle the position vector traces over such a curve must be zero.

7. Show that the field $\mathbf{F} = 2xy\mathbf{i} + (x^2 + z)\mathbf{j} + y\mathbf{k}$ is conservative.
   Solution: $\mathbf{F} = \text{grad}\phi$ where $\phi = x^2 y + yz$.

4.4 Conservative Fields: Irrotational Fields

1. Use the zero curl test to determine whether the following fields are conservative:

   (a) $\mathbf{F} = (12xy + yz)\mathbf{i} + (6x^2 + xz)\mathbf{j} + xy\mathbf{k}$
   (b) $\mathbf{F} = ze^{xz}\mathbf{i} + xe^{xz}\mathbf{k}$
   (c) $\mathbf{F} = \sin x \mathbf{i} + y^2 \mathbf{j} + e^z \mathbf{k}$
   (d) $\mathbf{F} = 3x^2 yz^2 \mathbf{i} + x^3 z^2 \mathbf{j} + x^3 yz \mathbf{k}$
   (e) $\mathbf{F} = \frac{2x}{x^2+y^2}\mathbf{i} + \frac{2y}{x^2+y^2}\mathbf{j} + 2z\mathbf{k}$

   For which of the fields is the test not applicable? How, then, can you test this field to determine whether it is conservative in its domain of influence?
   Solution: If $\mathbf{F}$ is the gradient of a potential,

   (a) Because the field $\mathbf{F}$ is defined everywhere, we can use the curl test. The curl is
   $\begin{vmatrix} \mathbf{i} & \mathbf{j} & \mathbf{k} \\ \partial_x & \partial_y & \partial_z \\ (12xy+yz) & (6x^2+xz) & xy \end{vmatrix} = 0$ Also, $\mathbf{F}$ is the gradient of a potential function $\phi = 6x^2 y + xyz + C$.

   (b) Because the field $\mathbf{F}$ is defined everywhere, we can use the curl test. The curl is
   $\begin{vmatrix} \mathbf{i} & \mathbf{j} & \mathbf{k} \\ \partial_x & \partial_y & \partial_z \\ ze^{xz} & 0 & xe^{xz} \end{vmatrix} = 0$ Also, $\mathbf{F}$ is the gradient of a potential function $\phi = e^{xz} + C$.

   (c) Because the field $\mathbf{F}$ is defined everywhere, we can use the curl test. The curl is
   $\begin{vmatrix} \mathbf{i} & \mathbf{j} & \mathbf{k} \\ \partial_x & \partial_y & \partial_z \\ \sin x & y^2 & e^z \end{vmatrix} = 0$. Also, $\mathbf{F}$ is the gradient of a potential function $\phi = -\cos x + \frac{1}{3}y^3 + e^z + C$.

4.4. CONSERVATIVE FIELDS: IRROTATIONAL FIELDS

(d) Because the field **F** is defined everywhere, we can use the curl test. The curl is
$$\begin{vmatrix} \mathbf{i} & \mathbf{j} & \mathbf{k} \\ \partial_x & \partial_y & \partial_z \\ 3x^2yz^2 & x^3z^2 & x^3yz \end{vmatrix} = (-x^3z)\mathbf{i} + (3x^2yz)\mathbf{j} + (0)\mathbf{k}, \text{ so the field is not conservative}$$

(e) Because the field $\mathbf{F} = \frac{2x}{x^2+y^2}\mathbf{i} + \frac{2y}{x^2+y^2}\mathbf{j} + 2z\mathbf{k}$ is not defined along the line $x = y = 0$, D is not simply connected and we cannot use the curl test. However, we can find the potential function by integrating: $\phi = \int \frac{2x}{x^2+y^2}dx = \ln(x^2 + y^2) + f(y, z)$, $\phi = \int \frac{2y}{x^2+y^2}dy = \ln(x^2 + y^2) + g(x, z)$ and $\phi = \int 2zdz = z^2 + h(x, y)$. Taking derivatives of the first with respect to y and z, we find $\phi_y = \frac{2y}{x^2+y^2} + f_y = F_2 = \frac{2y}{x^2+y^2}$, so $f = f(z)$, that is, f is a function of z alone. Also, $\phi_z = F_3 = 2z$, so $f = z^2$. Further similar integrations show that $\phi = \ln(x^2 + y^2) + z^2$.

3. Show that the scalar field $\phi = -1/|\mathbf{R}|$, which is defined everywhere except at the origin, is a potential function for the vector field $\mathbf{R}/|\mathbf{R}|^3$, where $\mathbf{R} = x\mathbf{i} + y\mathbf{j} + z\mathbf{k}$.

   (a) by writing ϕ in terms of x, y and z and computing its gradient.

   (b) by inspection, using the second and third properties of the gradient listed in section 3.1.

   Solution:

   (a) Because $\phi = \frac{-1}{\sqrt{x^2+y^2+z^2}}$, $\mathrm{grad}\phi = \frac{x}{(x^2+y^2+z^2)^{3/2}}\mathbf{i} + \frac{y}{(x^2+y^2+z^2)^{3/2}}\mathbf{j} + \frac{z}{(x^2+y^2+z^2)^{3/2}}\mathbf{k} = \frac{\mathbf{R}}{(\mathbf{R}^{1/2})^3} = \frac{\mathbf{R}}{|\mathbf{R}|^3}$.

   (b) Property 2 says that $\mathrm{grad}\phi$ points in the direction of the maximum rate of increase of ϕ. ϕ grows larger and larger negative as we approach the origin, and $\mathbf{R}/|\mathbf{R}|^3$ points away from the origin, which is in the direction of greatest increase. Property 3 says that the magnitude of $\mathrm{grad}\phi$ is equal to the maximum rate of increase of ϕ per unit distance, and because $\mathbf{R}/|\mathbf{R}|$ is a unit vector, the magnitude of $\mathbf{R}/|\mathbf{R}|^3$ is $1/|\mathbf{R}|^2$, which is the rate of increase of the function $-1/|\mathbf{R}|$.

5. Would your answer to exercise 4 be any different if the path extending from $(1, 2, 3)$ to $(2, 3, 5)$ were not straight?
   Solution: In exercise (4), $\mathbf{F} = \frac{x}{x^2+y^2+z^2}\mathbf{i} + \frac{y}{x^2+y^2+z^2}\mathbf{j} + \frac{z}{x^2+y^2+z^2}\mathbf{k}$, so we cannot use the curl test, as D is not simply connected. However, because $\mathbf{F}$ is the gradient of a potential, it is conservative and its integral over any path between $(1, 2, 3)$ and $(2, 3, 5)$ through D will be the same, as long as the path does not pass through the origin $(0, 0, 0)$ where $\mathbf{F}$ is not defined.

7. Let
$$\mathbf{F} = [(1+x)e^{x+y}]\mathbf{i} + [xe^{x+y} + 2y]\mathbf{j} - 2z\mathbf{k},$$
$$\mathbf{G} = [(1+x)e^{x+y}]\mathbf{i} + [xe^{x+y} + 2z]\mathbf{j} - 2y\mathbf{k}.$$

   (a) Show that **F** is conservative by finding a potential ϕ for **F**.

   (b) Evaluate $\int_C \mathbf{G} \cdot d\mathbf{R}$, where C is the path given by $x = (1-t)e^t$, $y = t$, $z = 2t$, $(0 \le t \le 1)$. (*Hint:* Take advantage of the similarity between **F** and **G**.)

   Solution:

   (a) If there is a potential ϕ, then the ith component of the vector is $\partial_i \phi$ and ϕ must be $\phi = \int (1+x)e^{x+y}dx = \int xe^{x+y} + 2y dy = \int -2z dz$, in which case $\phi = xe^{x+y} + f(y, z) + C = xe^{x+y} + y^2 + g(x, z) + C = -z^2 + h(x, y) + C$. Because we are integrating a function of several variables with respect to a single variable, the "constant" of integration is the sum of a function of the other two variables only and a constant. Comparing these, we can conclude that $\phi = xe^{x+y} + y^2 - z^2 + C$. To be sure, we can compare the components of **F** to the derivatives of the ϕs. For example, $\frac{\partial \phi}{\partial y} = xe^{x+y} + 2y = xe^{x+y} + f_y(y, z)$, so $f(y, z) = \int f_y(y, z)dy = \int 2y dy = y^2 + \hat{f}(z)$, while $\frac{\partial \phi}{\partial z} = -2z = f_z(y, z)$, so $f(y, z) = \int f_z(y, z)dz = \int -2z dy = -z^2 + \hat{f}(y)$. We conclude that $f(y, z) = y^2 - z^2$, so $\phi = xe^{x+y} + y^2 - z^2$, and we are done.

   (b) The trick here is to see that **F** and **G** are very much alike; in fact their difference is $\mathbf{F} - \mathbf{G} = (2y - 2z)\mathbf{j} + (2y - 2z)\mathbf{k}$, so $\mathbf{G} = \mathbf{F} - ((2y - 2z)\mathbf{j} + (2y - 2z)\mathbf{k})$. Therefore
$$\int_0^1 \mathbf{G} \cdot \frac{d\mathbf{R}}{dt} dt = \int_0^1 (\mathbf{F} - ((2y - 2z)\mathbf{j} + (2y - 2z)\mathbf{k})) \cdot \frac{d\mathbf{R}}{dt} dt =$$

$$= \int_0^1 \mathbf{F} \cdot \frac{d\mathbf{R}}{dt} dt - \int_0^1 ((2y-2z)\mathbf{j} + (2y-2z)\mathbf{k}) \cdot \frac{d\mathbf{R}}{dt} dt.$$

Because $\mathbf{F}$ is conservative, $\int_0^1 \nabla\phi \cdot \frac{d\mathbf{R}}{dt} dt = \phi(1) - \phi(0)$, which we can easily compute. At $t = 0$, $\mathbf{R} = (1-t)e^t\mathbf{i} + t\mathbf{j} + 2t\mathbf{k}$ points to $(1,0,0)$ and at $t = 1$, $\mathbf{R}$ points to $(0,1,2)$, so $\phi(t=1) - \phi(t=0) = \phi(0,1,2) - \phi(1,0,0) = -3-e$, where we have abused the notation just a little. Along the path traced by $\mathbf{R} = (1-t)e^t\mathbf{i} + t\mathbf{j} + 2t\mathbf{k}$, the vector field $((2y-2z)\mathbf{j} + (2y-2z)\mathbf{k} = ((2t-4t)\mathbf{j} + (2t-4t)\mathbf{k})) = -2t\mathbf{j} - 2t\mathbf{k}$, so we integrate $-2\int_0^1 (t\mathbf{j} + t\mathbf{k}) \cdot (-te^t\mathbf{i} + \mathbf{j} + 2\mathbf{k}) dt = -2\int_0^1 3t\, dt = -3$. Therefore $\int_0^1 \mathbf{G} \cdot \frac{d\mathbf{R}}{dt} dt = -3 - e - (-3) = -e$.

9. (a) Find a potential ϕ for the field

$$\mathbf{F} = (2xyz + z^2 - 2y^2 + 1)\mathbf{i} + (x^2z - 4xy)\mathbf{j} + (x^2y + 2xz - 2)\mathbf{k}$$

   (b) The field

$$\mathbf{G} = \frac{x}{(x^2+z^2)^2}\mathbf{i} + \frac{z}{(x^2+z^2)^2}\mathbf{k}$$

satisfies the condition that $\nabla \times \mathbf{G} = \mathbf{0}$ at all points except on the y axis. Is $\mathbf{G}$ conservative?

Solution:

(a) If $\mathbf{F}$ is the gradient of a potential ϕ, then $\nabla\phi = \mathbf{F} = \phi_x\mathbf{i} + \phi_y\mathbf{j} + \phi_z\mathbf{k} = (2xyz + z^2 - 2y^2 + 1)\mathbf{i} + (x^2z - 4xy)\mathbf{j} + (x^2y + 2xz - 2)\mathbf{k}$, so to find ϕ we integrate $\phi = \int \phi_x dx = \int \phi_y dy = \int \phi_z dz$. We find $\phi_1 = \int 2xyz + z^2 - 2y^2 + 1\, dx = x^2yz + xz^2 - 2xy^2 + x + f(y,z) + C$, $\phi_2 = \int x^2z - 4xy\, dy = x^2yz - 2y^2 + g(x,z) + C$ and $\phi_3 = \int x^2y + 2xz - 2\, dz = x^2yz + xz^2 - 2z + h(x,y) + C$, where we have labeled ϕ by which integral it came from. These must be all the same. While we can find ϕ from this by inspection and a little experiment, lets take the extra step and compute f, g, h explicitly. Because $\frac{\partial \phi}{\partial y} = x^2z - 4xy$ and $\frac{\partial \phi_1}{\partial y} = x^2z - 4xy + f_y(yz)$ must be equal, and $f(y,z)$ must be a function of z only, say $\hat{f}(z)$. Now compare $\frac{\partial \phi_1}{\partial z} = x^2y + 2xz + \hat{f}_z$ and $\frac{\partial \phi}{\partial z} = x^2y + 2xz - 2$; together these imply $\hat{f}_z = -2$, so $\hat{f} = -2z + C$. Therefore $\phi = x^2yz + xz^2 - 2xy^2 + x - 2z + C$.

(b) Because $\mathbf{G}$ is not defined on the y axis, in the space $\mathbb{R}^3$ minus the y axis the curl test is not guaranteed to work. However, one can find its potential function $\phi = \frac{-1}{2(x^2+z^2)} + C$ which is defined over $\mathbb{R}^3$ minus the y axis, so it is indeed conservative in this domain.

11. If $\mathbf{F}$ and $\mathbf{G}$ are conservative fields, is $\mathbf{F} \times \mathbf{G}$ necessarily conservative?
Solution: No, as we can see by constructing a counterexample: take $\phi_1 = xy$, $\phi_2 = yz$. Then their gradients are $\nabla\phi_1 = y\mathbf{i} + x\mathbf{j}$ and $\nabla\phi_2 = z\mathbf{j} + y\mathbf{k}$ with vector product $\nabla\phi_1 \times \nabla\phi_2 = xy\mathbf{i} + y^2\mathbf{j} + yz\mathbf{k}$. This has nonzero curl, so it is not a conservative vector field.

13. A function $f(x,y,z)$ is said to be *homogeneous of degree* k if $f(tx,ty,tz) = t^k f(x,y,z)$. Suppose that the components F_1, F_2, F_3 of the vector field $\mathbf{F}(x,y,z)$ are each homogeneous of degree k, and curl $\mathbf{F} = 0$. Prove

$$\mathbf{F} = \nabla\left(\frac{xF_1 + yF_2 + zF_3}{k+1}\right).$$

Solution: This is a tricky one. How do you go about proving this? The best approach is to write down what you are given in components, then see where you need to go and try to get various pieces. Because we have the condition that the curl must be zero, we know we must use $\left(\frac{\partial F_3}{\partial y} - \frac{\partial F_2}{\partial z}\right)\mathbf{i} + \left(\frac{\partial F_1}{\partial z} - \frac{\partial F_3}{\partial x}\right)\mathbf{j} + \left(\frac{\partial F_2}{\partial x} - \frac{\partial F_1}{\partial y}\right)\mathbf{k} = \mathbf{0}$ somewhere, and because the result includes a gradient, we suspect we will have to substitute partial derivatives in one variable for partial derivatives in another. Somehow, we must get a $k+1$ from somewhere. It is easy to see that we can produce k if we take a derivative of the defining equation for a homogeneous function of degree k with respect to t, so let's see what that gets us. Let $\mathbf{F} = \mathbf{F}(tx, ty, tz)$, and take the derivative of $\mathbf{F}(tx, ty, tz) = t^k \mathbf{F}(x,y,z)$ (we can do this because if $\mathbf{F}$ is homogeneous in its components, we can factor a t^k out of each). Then $\frac{d\mathbf{F}}{dt} = \frac{\partial \mathbf{F}}{\partial(tx)}\frac{d(tx)}{dt} + \frac{\partial \mathbf{F}}{\partial(ty)}\frac{d(ty)}{dt} + \frac{\partial \mathbf{F}}{\partial(tz)}\frac{d(tx)}{dt} = x\frac{\partial \mathbf{F}}{\partial(tx)} + y\frac{\partial \mathbf{F}}{\partial(ty)} + z\frac{\partial \mathbf{F}}{\partial(tz)} = kt^{k-1}\mathbf{F}(x,y,z)$. Now set $t = 1$, and we have $x\frac{\partial \mathbf{F}}{\partial x} + y\frac{\partial \mathbf{F}}{\partial y} + z\frac{\partial \mathbf{F}}{\partial z} = k\mathbf{F}(x,y,z)$. This looks promising, but we are not there yet. Let's examine a component of this vector function and a component of

4.5. VECTOR POTENTIALS AND SOLENOIDAL FIELDS

$\nabla\left(\frac{xF_1+yF_2+zF_3}{k+1}\right)$. The function we just computed has components $x\frac{\partial F_i}{\partial x}+y\frac{\partial F_i}{\partial y}+z\frac{\partial F_i}{\partial z}$ while the second, ignoring the $\frac{1}{k+1}$, has components $x\frac{\partial F_1}{\partial x_i}+y\frac{\partial F_2}{\partial x_i}+z\frac{\partial F_3}{\partial x_i}$, where x_i is x, y, or z, depending on the component. Here is where we use the fact that the curl is zero: because $\frac{\partial F_3}{\partial y}=\frac{\partial F_2}{\partial z}$, $\frac{\partial F_1}{\partial z}=\frac{\partial F_3}{\partial x}$ and $\frac{\partial F_2}{\partial x}=\frac{\partial F_1}{\partial y}$, we can make some useful substitutions. Let's compare

$(k+1)\mathbf{F}=\nabla(xF_1+yF_2+zF_3)$
$=\left(F_1+x\frac{\partial F_1}{\partial x}+y\frac{\partial F_2}{\partial x}+z\frac{\partial F_3}{\partial x}\right)\mathbf{i}+\left(x\frac{\partial F_1}{\partial y}+F_2+y\frac{\partial F_2}{\partial y}+z\frac{\partial F_3}{\partial z}\right)\mathbf{j}+\left(x\frac{\partial F_1}{\partial z}+y\frac{\partial F_2}{\partial z}+F_3+z\frac{\partial F_3}{\partial z}\right)\mathbf{k}$

and

$k\mathbf{F}=\left(x\frac{\partial F_1}{\partial x}+y\frac{\partial F_1}{\partial y}+z\frac{\partial F_1}{\partial z}\right)\mathbf{i}+\left(x\frac{\partial F_2}{\partial x}+y\frac{\partial F_2}{\partial y}+z\frac{\partial F_2}{\partial z}\right)\mathbf{j}+\left(x\frac{\partial F_3}{\partial x}+y\frac{\partial F_3}{\partial y}+z\frac{\partial F_3}{\partial z}\right)\mathbf{k}.$

Subtracting $\mathbf{F}$ from both sides of the first equation, and substituting $\frac{\partial F_3}{\partial y}=\frac{\partial F_2}{\partial z}$, $\frac{\partial F_1}{\partial z}=\frac{\partial F_3}{\partial x}$ and $\frac{\partial F_2}{\partial x}=\frac{\partial F_1}{\partial y}$ into the second, we see that we have identical equations $k\mathbf{F}=\left(x\frac{\partial F_1}{\partial x}+y\frac{\partial F_2}{\partial x}+z\frac{\partial F_3}{\partial x}\right)\mathbf{i}+\left(x\frac{\partial F_1}{\partial y}+y\frac{\partial F_2}{\partial y}+z\frac{\partial F_3}{\partial y}\right)\mathbf{j}+\left(x\frac{\partial F_1}{\partial z}+y\frac{\partial F_2}{\partial z}+z\frac{\partial F_3}{\partial z}\right)\mathbf{k}$. Now we see how to get to the final result. To get from $k\mathbf{F}=\left(x\frac{\partial F_1}{\partial x}+y\frac{\partial F_1}{\partial y}+z\frac{\partial F_1}{\partial z}\right)\mathbf{i}+\left(x\frac{\partial F_2}{\partial x}+y\frac{\partial F_2}{\partial y}+z\frac{\partial F_2}{\partial z}\right)\mathbf{j}+\left(x\frac{\partial F_3}{\partial x}+y\frac{\partial F_3}{\partial y}+z\frac{\partial F_3}{\partial z}\right)\mathbf{k}$ to our goal, make the substitutions to get $k\mathbf{F}=\left(x\frac{\partial F_1}{\partial x}+y\frac{\partial F_2}{\partial x}+z\frac{\partial F_3}{\partial x}\right)\mathbf{i}+\left(x\frac{\partial F_1}{\partial y}+y\frac{\partial F_2}{\partial y}+z\frac{\partial F_3}{\partial z}\right)\mathbf{j}+\left(x\frac{\partial F_1}{\partial z}+y\frac{\partial F_2}{\partial z}+z\frac{\partial F_3}{\partial z}\right)\mathbf{k}$, add $\mathbf{F}$ to each side and note that $\left(F_i+x\frac{\partial F_1}{\partial x_i}+y\frac{\partial F_2}{\partial x_i}+z\frac{\partial F_3}{\partial x_i}\right)=\frac{\partial}{\partial x_i}(xF_1+yF_2+zF_3)$. Therefore,

$$\mathbf{F}=\frac{1}{k+1}\left(\frac{\partial}{\partial x}(xF_1+yF_2+zF_3)\mathbf{i}+\frac{\partial}{\partial y}(xF_1+yF_2+zF_3)\mathbf{j}+\frac{\partial}{\partial z}(xF_1+yF_2+zF_3)\mathbf{k}\right)=$$

$$=\nabla\frac{xF_1+yF_2+zF_3}{k+1}.$$

4.5 Vector Potentials and Solenoidal Fields

1. Verify that $\mathbf{F}=\nabla\times\mathbf{G}$ in examples 4.9 and 4.10.
   Solution: In 4.9, $\mathbf{F}=\mathbf{A}=a_1\mathbf{i}+a_2\mathbf{j}+a_3\mathbf{k}$ is a constant vector field, and $\mathbf{G}=\frac{1}{2}\mathbf{A}\times\mathbf{R}$. Write $\mathbf{G}$ in components as $\mathbf{G}=\frac{1}{2}[(a_2z-a_3y)\mathbf{i}+(a_3x-a_1z)\mathbf{j}+(a_1y-a_2x)\mathbf{k}]$. Then

$$\nabla\times\mathbf{G}=\frac{1}{2}\begin{bmatrix}\mathbf{i} & \mathbf{j} & \mathbf{k}\\ \partial_x & \partial_y & \partial_z\\ a_2z-a_3y & a_3x-a_1z & a_1y-a_2x\end{bmatrix}=\frac{1}{2}[(a_1-(-a_1))\mathbf{i}+(a_2-(-a_2))\mathbf{j}+(a_3-(-a_3))\mathbf{k}].$$

In 4.10, $\mathbf{F}=\mathbf{A}\times\mathbf{R}=(a_2z-a_3y)\mathbf{i}+(a_3x-a_1z)\mathbf{j}+(a_1y-a_2x)\mathbf{k}$ is a constant vector field, and $\mathbf{G}=\frac{1}{3}(\mathbf{A}\times\mathbf{R})\times\mathbf{R}$. First, we compute

$$\mathbf{G}=\frac{1}{3}(\mathbf{A}\times\mathbf{R})\times\mathbf{R}=\frac{1}{3}\begin{bmatrix}\mathbf{i} & \mathbf{j} & \mathbf{k}\\ a_2z-a_3y & a_3x-a_1z & a_1y-a_2x\\ x & y & z\end{bmatrix}=$$

$$=\frac{1}{3}[-a_1(y^2+z^2)+a_2xy+a_3xz]\mathbf{i}+[a_1xy-a_2(x^2+z^2)+a_3yz]\mathbf{j}+[a_1xz+a_2yz-a_3(x^2+y^2)]\mathbf{k}.$$

Then

$$\nabla\times\mathbf{G}=$$

$$\frac{1}{3}\begin{bmatrix}\mathbf{i} & \mathbf{j} & \mathbf{k}\\ \partial_x & \partial_y & \partial_z\\ (-a_1(y^2+z^2)+a_2xy+a_3xz) & (a_1xy-a_2(x^2+z^2)+a_3yz) & (a_1xz+a_2yz-a_3(x^2+y^2))\end{bmatrix}$$

4.5. VECTOR POTENTIALS AND SOLENOIDAL FIELDS

$$= \frac{1}{3}[3(a_2z - a_3y)\mathbf{i} + 3(a_3x - a_1z)\mathbf{j} + 3(a_1y - a_2x)\mathbf{k}] = (a_2z - a_3y)\mathbf{i} + (a_3x - a_1z)\mathbf{j} + (a_1y - a_2x)\mathbf{k}.$$

3. Verify directly that $\mathbf{G} = (\mathbf{A} \times \mathbf{R}) \times \mathbf{R}/|\mathbf{R}|^2$ is a vector potential for $\mathbf{F} = \mathbf{A} \times \mathbf{R}/|\mathbf{R}|^2$. Convince yourself that that the corresponding flow line pattern is identical with that in figure 4.17.
   Solution: Rewrite $\mathbf{G} = (\mathbf{A} \times \mathbf{R}) \times \mathbf{R}/|\mathbf{R}|^2$ $\mathbf{G} = 1/|\mathbf{R}|^2[\mathbf{R}(\mathbf{A} \cdot \mathbf{R}) - \mathbf{A}|\mathbf{R}|^2]$ as using the vector identity $(\mathbf{A} \times \mathbf{B}) \times \mathbf{C} = \mathbf{B}(\mathbf{A} \cdot \mathbf{C}) - \mathbf{A}(\mathbf{B} \cdot \mathbf{C})$. Then $\mathbf{G} = 1/|\mathbf{R}|^2\mathbf{R}(\mathbf{A} \cdot \mathbf{R}) - \mathbf{A}$ and $\nabla \times \mathbf{G} = \nabla \times [1/|\mathbf{R}|^2\mathbf{R}(\mathbf{A} \cdot \mathbf{R}) - \mathbf{A}] = \nabla \times 1/|\mathbf{R}|^2\mathbf{R}(\mathbf{A} \cdot \mathbf{R})$ because $\mathbf{A}$ is constant. Now, because $\nabla \times \phi \mathbf{Q} = \nabla \phi \times \mathbf{Q} + \phi \nabla \times \mathbf{Q}$ for ϕ a scalar, we have $\nabla \times \mathbf{G} = \nabla [1/|\mathbf{R}|^2(\mathbf{A} \cdot \mathbf{R})] \times \mathbf{R} + [1/|\mathbf{R}|^2(\mathbf{A} \cdot \mathbf{R})]\nabla \times \mathbf{R}$. But $\nabla \times \mathbf{R} = 0$, so we have $\nabla \times \mathbf{G} = \nabla [1/|\mathbf{R}|^2(\mathbf{A} \cdot \mathbf{R})] \times \mathbf{R} = (-2\mathbf{R}(\mathbf{A} \cdot \mathbf{R})/|\mathbf{R}|^4 + \nabla(\mathbf{A} \cdot \mathbf{R})/|\mathbf{R}|^2) \times \mathbf{R} = \frac{1}{|\mathbf{R}|^2}(\mathbf{A} \times \mathbf{R})$ because $\mathbf{R} \times \mathbf{R} = 0$ and $\nabla(\mathbf{A} \cdot \mathbf{R}) = \mathbf{A}$.

5. Verify the claims made for various vector potentials $\mathbf{G}_1$, $\mathbf{G}_2$ and $\mathbf{G}_3$ for the rotating fluid field. Compute the divergence of each one.
   Solution: Given $\mathbf{A}$ constant and position vector $\mathbf{R}$, we must show

   (a) $\frac{1}{3}(\mathbf{A} \times \mathbf{R}) \times \mathbf{R} + \frac{1}{3}\nabla(|\mathbf{R}|^2\mathbf{A} \cdot \mathbf{R}) = (\mathbf{A} \cdot \mathbf{R})\mathbf{R}$

   (b) $\frac{1}{3}(\mathbf{A} \times \mathbf{R}) \times \mathbf{R} - \frac{1}{6}\nabla(|\mathbf{R}|^2\mathbf{A} \cdot \mathbf{R}) = -\frac{1}{2}(\mathbf{R} \cdot \mathbf{R})\mathbf{A}$

   To show the first, note that $(\mathbf{A} \times \mathbf{R}) \times \mathbf{R} = (\mathbf{A} \cdot \mathbf{R})\mathbf{R} - (\mathbf{R} \cdot \mathbf{R})\mathbf{A}$ (identity 1.31) and $\nabla(|\mathbf{R}|^2\mathbf{A} \cdot \mathbf{R}) = \nabla(\mathbf{R} \cdot \mathbf{R} \mathbf{A} \cdot \mathbf{R}) = (\mathbf{A} \cdot \mathbf{R})\nabla(\mathbf{R} \cdot \mathbf{R}) + (\mathbf{R} \cdot \mathbf{R})\nabla(\mathbf{A} \cdot \mathbf{R})$ (vector identity 3.27). Furthermore, by identities 3.35 and 3.39, $(\mathbf{A} \cdot \mathbf{R})\nabla(\mathbf{R} \cdot \mathbf{R}) + (\mathbf{R} \cdot \mathbf{R})\nabla(\mathbf{A} \cdot \mathbf{R}) = (\mathbf{A} \cdot \mathbf{R})[2(\mathbf{R} \cdot \nabla)\mathbf{R} + 2\mathbf{R} \times (\nabla \times \mathbf{R})] + (\mathbf{R} \cdot \mathbf{R})\mathbf{A} = 2(\mathbf{A} \cdot \mathbf{R})\mathbf{R} + (\mathbf{R} \cdot \mathbf{R})\mathbf{A}$, so $\frac{1}{3}(\mathbf{A} \times \mathbf{R}) \times \mathbf{R} + \frac{1}{3}\nabla(|\mathbf{R}|^2\mathbf{A} \cdot \mathbf{R}) = (\mathbf{A} \cdot \mathbf{R})\mathbf{R}$. The second equation $\frac{1}{3}(\mathbf{A} \times \mathbf{R}) \times \mathbf{R} - \frac{1}{6}\nabla(|\mathbf{R}|^2\mathbf{A} \cdot \mathbf{R}) = -\frac{1}{2}(\mathbf{R} \cdot \mathbf{R})\mathbf{A}$ follows by substitution of these same identities. The divergences are

   (a) $\nabla \cdot \mathbf{G}_1 = \nabla \cdot \left(\frac{1}{3}(\mathbf{A} \times \mathbf{R}) \times \mathbf{R}\right) = \frac{1}{3}(\nabla \cdot [(\mathbf{A} \cdot \mathbf{R})\mathbf{R}] - \nabla \cdot [(\mathbf{R} \cdot \mathbf{R})\mathbf{A}]) =$
   $= \frac{1}{3}(\mathbf{A} \cdot \mathbf{R}\nabla \cdot \mathbf{R} + \mathbf{R} \cdot \nabla(\mathbf{A} \cdot \mathbf{R}) - (\mathbf{R} \cdot \mathbf{R})\nabla \cdot \mathbf{A} - \mathbf{A} \cdot \nabla(\mathbf{R} \cdot \mathbf{R})) =$
   $= \frac{1}{3}(3\mathbf{A} \cdot \mathbf{R} + \mathbf{R} \cdot \mathbf{A} - 0 - 2\mathbf{A} \cdot \mathbf{R}) = \frac{2}{3}\mathbf{A} \cdot \mathbf{R}$.

   (b) $\nabla \cdot \mathbf{G}_2 = \nabla \cdot ((\mathbf{A} \cdot \mathbf{R}) \times \mathbf{R}) = \mathbf{A} \cdot \mathbf{R}\nabla \cdot \mathbf{R} + \mathbf{R} \cdot \nabla(\mathbf{A} \cdot \mathbf{R}) = 3\mathbf{A} \cdot \mathbf{R} + \mathbf{R} \cdot \mathbf{A} = 4\mathbf{A} \cdot \mathbf{R}$.

   (c) $\nabla \cdot \mathbf{G}_3 = \nabla \cdot \left(-\frac{1}{2}(\mathbf{R} \times \mathbf{R}) \times \mathbf{A}\right) = -\frac{1}{2}((\mathbf{R} \cdot \mathbf{R})\nabla \cdot \mathbf{A} + \mathbf{A} \cdot \nabla(\mathbf{R} \cdot \mathbf{R})) = -\frac{1}{2}(2\mathbf{A} \cdot \mathbf{R}) = -\mathbf{A} \cdot \mathbf{R}$.

7. Prove: if $\mathbf{F}$ and $\mathbf{G}$ are irrotational, then $\mathbf{F} \times \mathbf{G}$ is solenoidal. Can you find the vector potential for $\mathbf{F} \times \mathbf{G}$? (*Hint:* The problem is considerably easier if you have mastered tensor notation.)
   Solution: If $\mathbf{F}$ and $\mathbf{G}$ are irrotational then $\nabla \times \mathbf{F} = \nabla \times \mathbf{G} = 0$. To show $\mathbf{F} \times \mathbf{G}$ is solenoidal, that is, $\nabla \cdot (\mathbf{F} \times \mathbf{G}) = 0$, we prove $\nabla \cdot (\mathbf{F} \times \mathbf{G}) = \mathbf{G} \cdot (\nabla \times \mathbf{F}) - \mathbf{F} \cdot (\nabla \times \mathbf{G})$. Using tensor notation, $\nabla \cdot (\mathbf{F} \times \mathbf{G}) = \partial_i \epsilon_{ijk} \mathbf{F}_j \mathbf{G}_k = \epsilon_{ijk}(\partial_i \mathbf{F}_j)\mathbf{G}_k + \epsilon_{ijk}(\partial_i \mathbf{G}_k)\mathbf{F}_j = \mathbf{G}_k \epsilon_{kij}(\partial_i \mathbf{F}_j) - \mathbf{F}_j \epsilon_{jik}(\partial_i \mathbf{G}_k) = \mathbf{G} \cdot (\nabla \times \mathbf{F}) - \mathbf{F} \cdot (\nabla \times \mathbf{G})$. Because $\nabla \times \mathbf{F} = \nabla \times \mathbf{G} = 0$, the result follows.

9. Compute the vector potentials for the following two dimensional fields:

   (a) $\mathbf{F} = \mathbf{A} = $ constant

   (b) $\mathbf{F} = xy\mathbf{i} - y^2/2\mathbf{j}$

   (c) $[-y/(x^2 + y^2)]\mathbf{i} + [x/(x^2 + y^2)]\mathbf{j}$ (*Hint:* Consult exercise 4.)

   Solution:

   (a) To find the vector potential, we must first calculate $\mathbf{k} \times \mathbf{F}$. $\mathbf{k} \times \mathbf{F} = \begin{vmatrix} \mathbf{i} & \mathbf{j} & \mathbf{k} \\ 0 & 0 & 1 \\ A_1 & A_2 & 0 \end{vmatrix} = A_1\mathbf{j} - A_2\mathbf{i}$.
   Next we simply need to determine $\mathbf{G}$ such that $\nabla \times \mathbf{G} = A_1\mathbf{j} - A_2\mathbf{i}$. Since there is no $\mathbf{k}$ term we know that the solution must look like $0\mathbf{i} + 0\mathbf{j} + a\mathbf{k}$. So, taking the curl of $a\mathbf{k}$ we see that $\frac{\partial a}{\partial y} = A_1$ and $\frac{\partial a}{\partial x} = -A_2$. Thus, $\mathbf{G} = (A_1 y - A_2 x)\mathbf{k}$.

   (b) As above, we first calculate $\mathbf{k} \times \mathbf{F}$. $\mathbf{k} \times \mathbf{F} = \begin{vmatrix} \mathbf{i} & \mathbf{j} & \mathbf{k} \\ 0 & 0 & 1 \\ xy & -y^2/2 & 0 \end{vmatrix} = y^2/2\mathbf{i} + xy\mathbf{j}$. Next we determine $\mathbf{G}$ with the assumption that it is of the form $a\mathbf{k}$. Thus, $\frac{\partial a}{\partial x} = y^2/2$ and $\frac{\partial a}{\partial y} = xy$. Therefore $\mathbf{G} = \frac{xy^2}{2}\mathbf{k}$.

61

4.6. ORIENTED SURFACES

(c) As before, we begin by calculating $\mathbf{k} \times \mathbf{F}$. $\mathbf{k} \times \mathbf{F} = \begin{vmatrix} \mathbf{i} & \mathbf{j} & \mathbf{k} \\ 0 & 0 & 1 \\ -y/(x^2+y^2) & x/(x^2+y^2) & 0 \end{vmatrix} = -x/(x^2+y^2)\mathbf{i} - y/(x^2+y^2)\mathbf{j}$. Next we determine $\mathbf{G}$ with the assumption that it is of the form $a\mathbf{k}$. Thus, $\frac{\partial a}{\partial x} = -x/(x^2+y^2)$ and $\frac{\partial a}{\partial y} = -y/(x^2+y^2)$. We can solve these integration problems by looking at $\frac{\partial a}{\partial x} = x/(x^2+y^2)$. This can be rewritten as $a = -\int \frac{x}{x^2+y^2} dx$. Using the substitution $u = \sqrt{x^2+y^2}$ and $du = x/\sqrt{x^2+y^2}$, we see that $a = -\int 1/u\, du = -ln\, u$. So, in the original problem, $\mathbf{G} = -ln(x^2+y^2)\mathbf{k}$.

4.6 Oriented Surfaces

1. Find the elements of surface area $d\mathbf{S}$ and dS in terms of du and dv for the surface S given parametrically by $x = u^2$, $y = \sqrt{2}uv$ and $z = v^2$.
   Solution: The position vector is $\mathbf{R} = u^2\mathbf{i} + \sqrt{2}uv\mathbf{j} + v^2\mathbf{k}$, and tangent vectors in the direction of the coordinate curves are $\frac{\partial \mathbf{R}}{\partial u} = 2u\mathbf{i} + \sqrt{2}v\mathbf{j}$ and $\frac{\partial \mathbf{R}}{\partial v} = \sqrt{2}u\mathbf{j} + 2v\mathbf{k}$. The area element is the length of the vector product of these, times the "raw" area element $du\, dv$. The vector product is $\frac{\partial \mathbf{R}}{\partial u} \times \frac{\partial \mathbf{R}}{\partial v} = 2\sqrt{2}v^2\mathbf{i} - 4uv\mathbf{j} + 2\sqrt{2}u^2\mathbf{k}$ and its magnitude is $2\sqrt{2}(v^2 + u^2)$, so the area element is $2\sqrt{2}(v^2 + u^2)du\, dv$.

3. Determine the element of surface area dS for a right circular cylinder $x = a\cos u$, $y = a\sin u$, $z = v$. Interpret geometrically. (See fig. 4.27.)
   Solution: We have $\mathbf{R} = a\cos u\mathbf{i} + a\sin u\mathbf{k} + v\mathbf{k}$, and taking derivatives in the u and v directions gives tangent vectors to the coordinate curves: $\frac{\partial \mathbf{R}}{\partial u} = -a\sin u\mathbf{i} + a\cos u\mathbf{j}$ and $\frac{\partial \mathbf{R}}{\partial v} = \mathbf{k}$. The surface area element is the magnitude of the vector product of the tangent vectors to the coordinate curves times the infinitesimal length elements along the coordinate curves. Because $\frac{\partial \mathbf{R}}{\partial u} \times \frac{\partial \mathbf{R}}{\partial v} = a\cos u\mathbf{i} + a\sin u\mathbf{j}$ and $|a\cos u\mathbf{i} + a\sin u\mathbf{j}| = a$, the area element is $dS = a\, du\, dv$.

5. Find the area of the section of the surface $x = u^2$, $y = uv$, $z = \frac{1}{2}v^2$ bounded by the curves $u = 0$, $u = 1$, $v = 0$, and $v = 3$.
   Solution: As in problem 3, we find the area element by taking the magnitude of the vector product of tangent vectors to the coordinate curves. The position vector in this coordinate system is $\mathbf{R} = u^2\mathbf{i} + uv\mathbf{j} + \frac{1}{2}v^2\mathbf{k}$, so tangent vectors to coordinate curves are $\frac{\partial \mathbf{R}}{\partial u} = 2u\mathbf{i} + v\mathbf{j}$ and $\frac{\partial \mathbf{R}}{\partial v} = u\mathbf{j} + v\mathbf{k}$, whose vector product is $\frac{\partial \mathbf{R}}{\partial u} \times \frac{\partial \mathbf{R}}{\partial v} = v^2\mathbf{i} - 2uv\mathbf{j} + 2u^2\mathbf{k}$. The magnitude is $|v^2\mathbf{i} - 2uv\mathbf{j} + 2u^2\mathbf{k}| = \sqrt{v^4 + 4u^2v^2 + 4u^4} = v^2 + 2u^2$, so the area element is $dS = (v^2 + 2u^2)du\, dv$. The surface are is $\int_{u=0}^{1}\int_{v=0}^{3} dS = \int_{u=0}^{1}\int_{v=0}^{3}(v^2 + 2u^2)du\, dv = 11$.

7. Draw a diagram similar to those of figures 4.23 and 4.24 for the surface of a tetrahedron.
   Solution: See Figure 4.1.

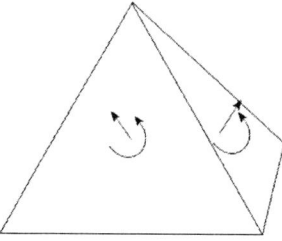

Figure 4.1: Orientations on the faces of a tetrahedron.

9. Show that the Möbius strip is piecewise smooth, and show why the smooth parts cannot be oriented consistently.
   Solution: A Möbius strip can be parameterized by $\mathbf{M} = (1 + \frac{v}{2}\cos\frac{u}{2})\cos u\mathbf{i} + (1 + \frac{v}{2}\cos\frac{u}{2})\sin u\mathbf{j} +$

$\frac{v}{2}\sin\frac{u}{2}\mathbf{k}$, which can be seen to be smooth (it has infinitely many derivatives). Now look at the form of the derivatives along the coordinate directions

$$\frac{\partial \mathbf{M}}{\partial u} = [-1/4\, v \sin(1/2\, u) \cos(u) - (1 + 1/2\, v \cos(1/2\, u)) \sin(u)]\,\mathbf{i}+$$

$$+ [-1/4\, v \sin(1/2\, u) \sin(u) + (1 + 1/2\, v \cos(1/2\, u)) \cos(u)]\,\mathbf{j}+$$

$$+\frac{1}{4} v \cos(1/2\, u)\,\mathbf{k}$$

and

$$\frac{\partial \mathbf{M}}{\partial v} = 1/2 \cos(1/2\, u) \cos(u)\,\mathbf{i} + 1/2 \cos(1/2\, u) \sin(u)\,\mathbf{j} + 1/2 \sin(1/2\, u)\,\mathbf{k}$$

with vector product

$$\frac{\partial \mathbf{M}}{\partial u} \times \frac{\partial \mathbf{M}}{\partial v} = 1/2 \sin(1/2\, u) \left(-1 + v (\cos(1/2\, u))^3 + 2 (\cos(1/2\, u))^2 - v \cos(1/2\, u)\right)\mathbf{i}+$$

$$+ \left[3/4\, v (\cos(1/2\, u))^2 - 1/2 (\cos(1/2\, u))^4 v + \cos(1/2\, u) - 1/8\, v - (\cos(1/2\, u))^3\right]\mathbf{j}$$

$$-1/4(\cos(1/2\, u)(2 + v \cos(1/2\, u)))\mathbf{k}.$$

When we substitute $v = 0$ and $u = 0$ into this we get $\frac{1}{2}\mathbf{k}$, which is the "outward" normal at the center of the strip, and for $v = 0$ and $u = 2\pi$ we get $-\frac{1}{2}\mathbf{k}$. If the surface was orientable, these should point in the same direction, because at these values of u and v, we are on the same point in the surface $\mathbf{M}$, which is $\mathbf{M}(0,0) = \mathbf{i}$.

4.7 Surface Integrals

1. If $\mathbf{F} = z\mathbf{k}$, find the surface integral of the normal component of $\mathbf{F}$ over the closed surface of the right circular cylinder with curved surface $x^2 + y^2 = 9$ and bases in the planes $z = 0$ and $z = 2$. (Mental arithmetic should suffice.)
   Solution: Because the vector field is at right angles to the normal to the sides of the cylinder, the integral will be zero. The integral over the disc at the bottom of the cylinder is zero because the vector field is zero there, and the integral over the top is two times the area, or 18π because the vector field dotted into the normal has value 2 there.

3. Compute the surface integral of the normal component of $\mathbf{F} = x\mathbf{i}$ over the triangle with vertices $(1,0,0)$, $(0,2,0)$, $(0,0,3)$, taking the normal on the side away from the origin.
   Solution: Two vectors in the plane of the triangle are $(1,0,0) - (0,0,3) = \mathbf{i} - 3\mathbf{k}$ and $(0,2,0) - (0,0,3) = 2\mathbf{j} - 3\mathbf{k}$ and their cross product is $6\mathbf{i} + 3\mathbf{j} + 2\mathbf{k}$ or unitized $\mathbf{n} = 1/7(6\mathbf{i} + 3\mathbf{j} + 2\mathbf{k})$. Because it is difficult to get appropriate limits for a parameterized non-planar triangle (without using linear interpolation, which has not been covered), we use the projection method. The integral is $\int_{x=0}^{x=1} \int_{y=0}^{y=2-2x} (x\mathbf{i}) \cdot (\frac{6}{7}\mathbf{i} + \frac{3}{7}\mathbf{j} + \frac{2}{7}\mathbf{k}) \frac{x}{\cos\gamma} dy\,dx = \int_{x=0}^{x=1} \int_{y=0}^{y=2-2x} \left(\frac{6}{7}\right) \frac{x}{\cos\gamma} dy\,dx$, where the $\frac{1}{\cos\gamma}$ is the factor that scales the area of the triangle to that projected on the plane. This factor is the dot product of the outward normal to the surface with the unit vector perpendicular to the plane of the projection, which in this case is the xy plane. Thus we can calculate $\cos\gamma = \mathbf{n} \cdot \mathbf{k} = 2/7$. Therefore we have $\int_{x=0}^{x=1} \int_{y=0}^{y=2-2x} \left(\frac{6}{7}\right) x \left(\frac{7}{2}\right) dy\,dx = 1$

5. Given $\mathbf{F} = x\mathbf{i} + y\mathbf{j} + (z^2 - 1)\mathbf{k}$ find $\iint \mathbf{F} \cdot \mathbf{n}\,dS$ over the closed surface bounded by the planes $z = 0$, $z = 1$, and the cylinder $x^2 + y^2 = a^2$, where $\mathbf{n}$ is the outward unit normal.
   Solution: The integral over the surface of the cylinder is $2\pi a^2$, the integral over the top surface is 0 because $z^2 - 1$ is 0 and the integral over the bottom disk is $a^2\pi$ because $z^2 - 1 = 1$ there. The total is $3a^2\pi$.

7. Let D be the region $x \geq 0$, $y \geq 0$, $z \geq 0$, $x + \frac{1}{2}y + \frac{1}{3}z \leq 1$.

4.7. SURFACE INTEGRALS

(a) Is the region a domain?

(b) Is the region simply connected?

(c) If $\mathbf{F} = 2x\mathbf{i} + y\mathbf{j} + z\mathbf{k}$, find the surface integral of the normal component of $\mathbf{F}$ over the boundary of this region, oriented by selecting the outward normal.

Solution:

(a) By definition, *domain* is a region that is both open and connected, so because this region includes some boundaries, which are closed, it cannot be open, so it is not a domain.

(b) The region described is the region in the first octant below a plane, so it is open and connected with no obstructions, and thus is simply connected.

(c) On the three sides of the tetrahedron bounded by the xz, yz and xy coordinate planes, the outward normal unit vectors are $-\mathbf{i}$, $-\mathbf{j}$ and $-\mathbf{k}$ respectively, so we must integrate $-\mathbf{i} \cdot (2x\mathbf{i} + y\mathbf{j} + z\mathbf{k}) = -2x$, $-\mathbf{j} \cdot (2x\mathbf{i} + y\mathbf{j} + z\mathbf{k}) = -y$, and $-\mathbf{k} \cdot (2x\mathbf{i} + y\mathbf{j} + z\mathbf{k}) = -z$, over the planes $x = 0$, $y = 0$ and $z = 0$. It is easy to see that the integrals will be zero in all three cases. The projection of the final surface on the xy plane is bounded by $x = 0$, $y = 0$ and $y = 2 - 2x$. Writing the formula for the plane in standard form $x + \frac{1}{2}y + \frac{1}{3}z = 1$, we can read off the normal to the plane $\mathbf{i} + \frac{1}{2}\mathbf{j} + \frac{1}{3}\mathbf{k}$, and its projection on $\mathbf{k}$ is $\cos\gamma = \frac{1}{3}$. Using the projection formula, we integrate $\int_{x=0}^{x=1} \int_{y=0}^{y=2-2x} \frac{\mathbf{F}\cdot\mathbf{n}}{\cos\gamma} dy dx = \int_{x=0}^{x=1} \int_{y=0}^{y=2-2x} \frac{(2x\mathbf{i}+y\mathbf{j}+z\mathbf{k})\cdot(\mathbf{i}+\frac{1}{2}\mathbf{j}+\frac{1}{3}\mathbf{k})}{\frac{1}{3}} dy dx = 3\int_{x=0}^{x=1} \int_{y=0}^{y=2-2x} 1 + x\, dy dx = 3\int_{x=0}^{x=1}(2-2x)(1+x)dx = 3\int_{x=0}^{x=1} 2 - 2x^2 dx = 4$.

9. Evaluate $\iint_S z^2 dS$, where S is the part of the lateral surface of the right circular cylinder $x^2 + y^2 = 4$ between the planes $z = 0$ and $z = y + 3$.
   Solution: The integral we wish to evaluate is $\int_{\theta=0}^{2\pi} \int_{z=0}^{z=y+3} z^2 dz r d\theta$ in cylindrical coordinates. Noting that $r = 2$ and $y = 2\sin\theta$, this becomes $\int_{\theta=0}^{2\pi} \int_{z=0}^{z=2\sin\theta+3} z^2 dz\, 2 d\theta$. Integrating once, we obtain $\int_{\theta=0}^{2\pi} \frac{1}{3}(2\sin\theta+3)^3 2d\theta\, 60\pi$ or $\frac{2}{3}\int_{\theta=0}^{2\pi}(2\sin\theta+3)^3 d\theta$. Integrating $(2\sin\theta+3)^3 = 8(\sin(\theta))^3 + 36(\sin(\theta))^2 + 54\sin(\theta) + 27$, we obtain $-8/3(\sin(\theta))^2\cos(\theta) - \frac{178}{3}\cos(\theta) - 18\cos(\theta)\sin(\theta) + 45\theta$, every term of which vanishes when evaluated from 0 to 2π except the 45θ. We find is $\frac{2}{3}(45\theta)|_0^{2\pi} = 60\pi$.

11. Our vector field is $\mathbf{F} = x^2\mathbf{i} + y^2\mathbf{j} + z^2\mathbf{k}$, and the surface S is the part of the cone $z^2 = x^2 + y^2$ for which $1 \le z \le 2$, with $\mathbf{n} \cdot \mathbf{k}$ positive.
    Solution: To begin we need to parameterize the surface of the cone between $z = 1$ and $z = 2$. This can be done a number of ways. However, the simplest is to parameterize between two circles.
    $C_1 = \cos t\mathbf{i} + \sin t\mathbf{j} + \mathbf{k}$
    $C_2 = 2\cos t\mathbf{i} + 2\sin t\mathbf{j} + 2\mathbf{k}$
    A linear interpolation between C_1 and C_2 gives the surface.
    $P = sC_1 + (1-s)C_2$
    $= (1+s)(\cos t\mathbf{i} + \sin t\mathbf{j} + \mathbf{k})$
    Next, we need to find the outward normal to the surface. To do this we begin by taking derivatives with respect to both s and t of P.
    $\frac{\partial P}{\partial s} = \cos t\mathbf{i} + \sin t\mathbf{j} + \mathbf{k}$
    $\frac{\partial P}{\partial t} = (1+s)(\sin t\mathbf{i} + \cos t\mathbf{j})$
    The outward normal to the surface is the cross of the two previously calculated vectors,

$$\mathbf{n} = \frac{\partial P}{\partial t} \times \frac{\partial P}{\partial s} = \begin{vmatrix} \mathbf{i} & \mathbf{j} & \mathbf{k} \\ \cos t & \sin t & 1 \\ -(1+s)\sin t & (1+s)\cos t & 0 \end{vmatrix} = (1+s)(-\cos t\mathbf{i} - \sin t\mathbf{j} + \mathbf{k})$$

The last step, before calculating the surface integral is to transform the initial $\mathbf{F}$ into the new parameterized coordinates.

$$\mathbf{F} = x^2\mathbf{i} + y^2\mathbf{j} + z^2\mathbf{k} = (1+s)^2\cos^2 t\mathbf{i} + (1+s)^2\sin^2 t\mathbf{j} + (1+s)^2\mathbf{k}$$

The final surface integral is

$$\iint \mathbf{F} \cdot \mathbf{n} = \int_0^{2\pi} \int_0^1 (1+s)^3(\cos^3 t + \sin^3 t + 1) ds dt = \int_0^{2\pi} \frac{15}{4}(\cos^3 t + \sin^3 t + 1) dt = \frac{15\pi}{2}.$$

13. $\mathbf{F} = y^2\mathbf{i} + z\mathbf{j} - x\mathbf{k}$; S is the part of the cylinder generated by $y^2 = 1 - x$ between the planes $z = 0$ and $z = x$ for $x \geq 0$ with $\mathbf{n} \cdot \mathbf{i} > 0$.
   Solution: The integral we want to evaluate is $\int\int \mathbf{F} \cdot d\mathbf{S}$. Because the surface is described by $x = 1 - y^2$ or $\mathbf{S} = (1 - y^2)\mathbf{i} + y\mathbf{j} + z\mathbf{k}$, we can compute the vector area element by calculating $\frac{\partial \mathbf{S}}{\partial y} \times \frac{\partial \mathbf{S}}{\partial z} = \mathbf{i} + 2y\mathbf{j}$, so $d\mathbf{S} = (\mathbf{i} + 2y\mathbf{j})dzdy$. Along this surface, $\mathbf{F} = y^2\mathbf{i} + z\mathbf{j} - (1-y^2)\mathbf{k}$. Then the integral is $\int_{y=-1}^{y=1} \int_{z=0}^{z=1-y^2} (y^2\mathbf{i} + z\mathbf{j} - (1-y^2)\mathbf{k}) \cdot (\mathbf{i}+2y\mathbf{j})dzdy = \int_{y=-1}^{y=1}\int_{z=0}^{z=1-y^2} y^2 + 2yz\, dz\, dy = 4/15$.

15. Let $\mathbf{E} = -\mathbf{grad}(|\mathbf{R}|^{-1})$, where $\mathbf{R} = x\mathbf{i} + y\mathbf{j} + z\mathbf{k}$.

    (a) Show that $\mathbf{E} = \mathbf{R}/|\mathbf{R}|^3$.

    (b) Find $\int_C \mathbf{E} \cdot d\mathbf{R}$, when C is the line segment joining the points $(0,1,0)$ and $(0,0,1)$.

    (c) Compute $\int\int_{S_1} \mathbf{E} \cdot d\mathbf{S}$, when S_1 is the sphere $x^2 + y^2 + z^2 = 9$.

    Solution:

    (a) Because $|\mathbf{R}|^{-1} = \frac{1}{\sqrt{x^2+y^2+z^2}}$, $\mathbf{E} = -\mathbf{grad}(|\mathbf{R}|^{-1}) = \left(\frac{x}{(x^2+y^2+z^2)^{3/2}}\right)\mathbf{i} + \left(\frac{y}{(x^2+y^2+z^2)^{3/2}}\right)\mathbf{j} + \left(\frac{z}{(x^2+y^2+z^2)^{3/2}}\right)\mathbf{k} = \frac{\mathbf{R}}{|\mathbf{R}|^3}$.

    (b) For the segment joining $(0,1,0)$ and $(0,0,1)$, $\mathbf{R} = (1-t)\mathbf{j} + t\mathbf{k}$, for $0 \leq t \leq 1$ and along this path $\mathbf{E} = \left(\frac{1-t}{((1-t)^2+t^2)^{3/2}}\right)\mathbf{j} + \left(\frac{t}{((1-t)^2+t^2)^{3/2}}\right)\mathbf{k} = \frac{1}{(1-2t+2t^2)^{3/2}}$. Integrating this from $t=0$ to $t=1$ we find $\int_{t=0}^{1} \frac{dt}{(1-2t+2t^2)^{3/2}} = \frac{2t-1}{(1-2t+2t^2)^{1/2}}\Big|_{t=0}^{1} = 2$.

    (c) In spherical coordinates $\mathbf{E} = -\mathbf{grad}(|\mathbf{R}|^{-1}) = -\mathbf{grad}\frac{1}{r} = \frac{1}{r^2}\mathbf{e}_r$. The outward unit normal to the sphere is $\mathbf{e}_r$, so we must integrate $\int_{\theta=0}^{\theta=2\pi}\int_{\phi=0}^{\phi=\pi} \frac{1}{r^2}\mathbf{e}_r \cdot \mathbf{e}_r r^2 \sin\phi d\phi d\theta = \int_{\theta=0}^{\theta=2\pi}\int_{\phi=0}^{\phi=\pi}\sin\phi d\phi d\theta = 4\pi$. Note that this is the area of the sphere divided by its radius squared, or the area of a unit sphere.

17. Consider a cone with vertex at the origin, as in figure 4.42. The *solid angle* Ω at the vertex is defined to be the surface area that this cone cuts out of the unit sphere centered at the origin.

    (a) If the cone is perfectly flat, that is, a plane, what is Ω?

    (b) What is Ω for the corner of the cube?

    (c) What is Ω for the $\phi = 45°$ cone?

    (d) What is the total solid angle around a point?

    (e) Suppose the surface S bounded by the simple closed curve C has the property that every ray from the origin intersects S at most once. Then the *solid angle Ω subtended at the origin by S* is the solid angle at the vertex of the cone generated by the rays through C. Show that if S is properly oriented,

    $$\Omega = \int\int_S \frac{\mathbf{R} \cdot d\mathbf{S}}{|\mathbf{R}|^3}, \quad (\mathbf{R} = x\mathbf{i} + y\mathbf{j} + z\mathbf{k}).$$

    Use the results to check Gauss' law, eq. (4.38), for a point charge at the origin. The expression for the electric field appears in example 4.25.

    Solution:

    (a) If the cone is perfectly flat, $\Omega = 2\pi$, or one half the area of the unit sphere.

    (b) Take a cube at the origin with its corners in the unit sphere. Then the cosine of the angle between a normal at the center of one of its faces and a ray from its center to a corner is $\frac{\sqrt{3}}{3}$. Therefore $\Omega = \int_0^{2\pi}\int_0^{\cos^{-1}\frac{\sqrt{3}}{3}} 1^2 \sin\phi d\phi d\theta = 2\pi(-\cos(\cos^{-1}\frac{\sqrt{3}}{3}) - (-\cos 0)) = 2\pi(1 - \frac{\sqrt{3}}{3})$.
    Note: The solid angle has units of *steradians*, and a cube subtends a solid angle of $1/6$ of the area of a unit sphere for each of its faces.

    (c) For the $\phi = 45°$ cone $\Omega = \int_0^{2\pi}\int_0^{\pi/4} 1^2 \sin\phi d\phi d\theta = 2\pi(1 - \frac{\sqrt{2}}{2})$.

    (d) Around a point the total solid angle is zero.

4.8. VOLUME INTEGRALS

(e) By definition the solid angle ω is the surface area cut out by the cone of the angle on the surface of a unit sphere centered at the origin. Therefore, the area of the projection of the surface S onto the surface of the unit sphere will give the solid angle. From problem 15 of this section, we know that the integral of the negative gradient of $\frac{1}{|\mathbf{R}|}$ over the surface of a sphere is equal to 4π divided by the radius squared, or equivalently, the area of the surface of a unit sphere. The negative gradient of $\frac{1}{|\mathbf{R}|}$ is $-\nabla \frac{1}{|\mathbf{R}|} = \frac{\mathbf{R}}{|\mathbf{R}|^3}$. Notice that in the integral in problem 15, we are just integrating a constant 1: there is no function that varies over the surface. Therefore, $\int \int_S \frac{\mathbf{R}}{|\mathbf{R}|^3} dS$ gives the same value as the area integral of the projection of S on the surface of the unit sphere, which is the solid angle Ω.

Gauss' law states that if S is a closed surface, the integral of the electrostatic field $\mathbf{E}$ generated by a charge q enclosed by the surface is $\int \int_S \mathbf{n} \cdot \mathbf{E} dS = \frac{q}{\varepsilon_0}$. The electrostatic field for a point charge at the origin is given by $\frac{q}{\varepsilon_0} \frac{\mathbf{R}}{|\mathbf{R}|^3}$, and substituting this into the integral $\int \int_S \mathbf{n} \cdot \frac{q}{\varepsilon_0} \frac{\mathbf{R}}{|\mathbf{R}|^3} dS = \frac{q}{\varepsilon_0}$ gives us the result we seek.

19. Consider a hollow sphere of homogeneous material, with an inner radius a and an outer radius b, with inner temperature T_a and outer temperature T_b.

    (a) Find the steady state temperature as a function of the distance r from the center, for values of r between a and b.

    (b) For a value of r halfway between a and b, is T halfway between T_a and T_b?

    Solution:

    (a) By Fourier's law, $\mathbf{Q} \cdot \mathbf{n} = (-k\nabla T) \cdot \mathbf{n}$, that is, the heat flow vector $\mathbf{Q}$ is in the opposite direction of the gradient of temperature T (heat flows from high temperature to low). We will integrate the flux of the heat flow vector over the surface of the sphere to determine the heat flow per unit time (any of the spherical surfaces will do, because what flows through one must flow through all the others. So calculate $H = \int \int \mathbf{Q} \cdot \mathbf{n} dS = \int \int -k\nabla T \cdot \mathbf{n} dS$. Because the unit normal vector is of the form $0\mathbf{e}_\theta + 0\mathbf{e}_\phi + 1\mathbf{e}_r$ and ∇T is of the form $\mathbf{e}_\theta + 0\mathbf{e}_\phi + \frac{\partial T}{\partial r}\mathbf{e}_r$, we have $\mathbf{n} \cdot \nabla T = \frac{\partial T}{\partial r}$. This is constant and may be taken outside of the integral to obtain $H = -k\frac{\partial T}{\partial r} \int \int dS = -4k\pi r^2 \frac{\partial T}{\partial r}$.

    Now we have a differential equation $H \int \frac{1}{r^2} dr = -4k\pi \int dT$ to solve to get the total heat flow per unit time. By taking the definite integral on both sides, we can solve for H, so integrate $H \int_{r=a}^{b} \frac{1}{r^2} dr = -4k\pi \int_{T=T_a}^{T_b} dT$ to get $H = 4\pi k \frac{T_b - T_a}{1/b - 1/a}$. Now we can solve for T in terms of r by substituting this expression for H into $H = -4k\pi r^2 \frac{\partial T}{\partial r}$ and integrating $4\pi k \frac{T_b - T_a}{1/b - 1/a} \int_{r=a}^{r} dr = -4k\pi \int_{T_a}^{T} dT$ to get $T = T_a + \frac{1/r - 1/a}{1/b - 1/a}(T_b - T_a)$.

    (b) T is not halfway between T_a and T_b: halfway between is $\frac{aT_a + bT_b}{a+b}$.

21. Compute the area of the cone $\phi =$ constant $= \pi/6$, $0 \leq r \leq 2$. [Hint: Use eqs. (4.25) and (3.52).]
    Solution: We can write a parameterization in spherical coordinates $\mathbf{R} = r\sin\phi\cos\theta\mathbf{i} + r\sin\phi\sin\theta\mathbf{j} + r\cos\phi$, which at $\phi = \pi/6$ is $\mathbf{R} = \frac{1}{2}r\cos\theta\mathbf{i} + \frac{1}{2}r\sin\theta\mathbf{j} + \frac{\sqrt{3}}{2}r\mathbf{k}$. The vectors in the direction of the coordinate directions are $\frac{\partial \mathbf{R}}{\partial r} = \frac{1}{2}\cos\theta\mathbf{i} + \frac{1}{2}\sin\theta\mathbf{j} + \frac{\sqrt{3}}{2}\mathbf{k}$ and $\frac{\partial \mathbf{R}}{\partial \theta} = -\frac{1}{2}r\sin\theta\mathbf{i} + \frac{1}{2}r\cos\theta\mathbf{j}$ and the area element is $|\frac{\partial \mathbf{R}}{\partial \theta} \times \frac{\partial \mathbf{R}}{\partial r}|drd\theta = |\frac{\sqrt{3}}{4}r\cos\theta\mathbf{i} + \frac{\sqrt{3}}{4}r\sin\theta\mathbf{j} - \frac{1}{4}r\mathbf{k}|drd\theta = \frac{r}{2}drd\theta$, so the area is $\int_0^{2\pi} \int_0^2 \frac{r}{2}drd\theta = \pi \int_0^2 rdr = 2\pi$.

4.8 Volume Integrals

1. Compute the volume of the sphere of radius R by iterated integrals.
   Solution: The volume is $\int_0^R \int_0^\pi \int_0^{2\pi} r^2 \sin\phi d\theta d\phi dr = 2\pi \int_0^R \int_0^\pi r^2 \sin\phi d\phi dr = 2\pi \frac{1}{3} R^3 \int_0^\pi \sin\phi d\phi = \frac{2}{3}\pi R^3 (-\cos\phi)|_0^\pi = 4/3\pi R^3$.

4.9. INTRODUCTION TO THE DIVERGENCE THEOREM AND STOKES' THEOREM

3. Sketch the region whose volume is represented by the triple integral

$$\int_0^2 \int_0^3 \int_0^{\sqrt{9-y^2}} dx\, dy\, dz.$$

Solution:

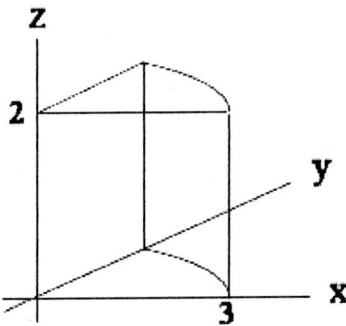

Figure 4.2: A cylinder of radius 3 and height 2 symmetric about the z axis and bounded by the planes $z = 0$ and $z = 2$.

5. Let V be a domain with volume η. Let $\mathbf{F} = x\mathbf{i} + y\mathbf{j} + z\mathbf{k}$.

(a) What is $\iiint_V \nabla \cdot \mathbf{F}\, dV$?

(b) On the basis of your answer to exercise 4, what do you conjecture is the value of $\int \in_S \mathbf{F} \cdot d\mathbf{S}$, the surface integral of the normal component of $\mathbf{F}$ over the boundary of V?

Solution:

(a) Because $\nabla \cdot \mathbf{F} = 3$, the integral of this quantity over a volume will be three times the volume.

(b) A reasonable conjecture, based on the fact that several different functions integrated over the same domain produced the same results, whether the integral was of the divergence of **G** over a volume or a surface integral of the outward normal component of **G** over a surface, would be that the integral of $\mathbf{F} = x\mathbf{i} + y\mathbf{j} + z\mathbf{k}$ dotted into the outward normal vector over a surface would be equal to 3 times the volume the surface encloses. However, just because it worked for a cube is not sufficient to claim it works for any surface. However, we can make the argument more plausible by noting that we can approximate any reasonable volume in space by a multitude of small cubes, and that the integrals of $\mathbf{F} \cdot \mathbf{n}$ over the interior faces cancel for pairs of cubes sharing the same face, because the outward normals for the shared faces are in opposite directions.

4.9 Introduction to the Divergence Theorem and Stokes' Theorem

1. Use the divergence theorem to solve exercise 1, section 4.7.
Solution: Exercise 1 is this: Given $\mathbf{F} = z\mathbf{k}$ find the flux of **F** over the cylinder $x^2 + y^2 = 9$ bounded by $z = 0$ and $z = 2$. The divergence theorem states that $\iiint \nabla \cdot \mathbf{F}\, dV = \iint \mathbf{F} \cdot \mathbf{n}\, dS$. In this case we have $\iiint \nabla \cdot z\mathbf{k}\, dV = \iiint 1\, dV$, so the integral is equal to the volume of the cylinder $V = bh = 9\pi \cdot 2 = 18\pi$.

3. Use the divergence theorem to solve

(a) Exercise 4, section 4.7.

(b) Exercise 5, section 4.7.

Solution:

4.9. INTRODUCTION TO THE DIVERGENCE THEOREM AND STOKES' THEOREM

(a) Exercise 4 states: Given $\mathbf{F} = x\mathbf{i} - y\mathbf{j}$, find the value of $\int\int \mathbf{F} \cdot \mathbf{n}dS$ over the closed surface bounded by the planes $z = 0$ and $z = 1$, and the cylinder $x^2 + y^2 = a^2$, where $\mathbf{n}$ is the unit outward normal. The divergence theorem states that $\int\int\int \nabla \cdot \mathbf{F}dV = \int\int \mathbf{F} \cdot \mathbf{n}dS$, so we calculate the divergence $\int\int\int \nabla \cdot (x\mathbf{i} - y\mathbf{j})dV = \int\int\int 0 dV = 0$, and we are done. Thinking of this geometrically, the vector field has no z component so the flux through the top and bottom of the cylinder must be zero, and the vector field is parallel to the sides of the cylinder, so again the flux must be zero there. It is more surprising that the flux of this vector field must be zero through any surface.

(b) Exercise 5 states: Given $\mathbf{F} = x\mathbf{i} + y\mathbf{j} + (z^2 - 1)\mathbf{k}$ find $\int\int \mathbf{F} \cdot \mathbf{n}dS$ over the closed surface bounded by the planes $z = 0$, $z = 1$, and the cylinder $x^2 + y^2 = a^2$, where $\mathbf{n}$ is the outward unit normal. Using the divergence $\int\int\int \nabla \cdot (x\mathbf{i} + y\mathbf{j} + (z^2-1)\mathbf{k})dV = \int\int\int \nabla \cdot (2+2z)dV$. Putting this in cylindrical coordinates $\int_{r=0}^{a}\int_{\theta=0}^{2\pi}\int_{z=0}^{1}(2+2z)\,dz\,d\theta\,r\,dr = \int_{r=0}^{a}\int_{\theta=0}^{2\pi}3\,d\theta\,r\,dr = 6\pi\int_{r=0}^{a}r\,dr = 3\pi a^2$.

5. Use the divergence theorem to evaluate $\int\int_S \mathbf{F} \cdot \mathbf{n}dS$, when $\mathbf{F} = y^2x\mathbf{i} + x^2y\mathbf{j} + z^2\mathbf{k}$, and when S is the complete surface of the region bounded by the cylinder $x^2 + y^2 = 4$ and by the planes $z = 0$ and $z = 2$.
   Solution: Using the divergence $\int\int\int \nabla \cdot \mathbf{F}dV = \int\int\int \nabla \cdot (y^2x\mathbf{i} + x^2y\mathbf{j} + z^2\mathbf{k})dV = \int\int\int (y^2 + x^2 + 2z)dV$. In cylindrical coordinates this integral is $\int_{\theta=0}^{2\pi}\int_{r=0}^{2}\int_{z=0}^{2}(r^2\cos^2\theta + r^2\sin^2\theta + 2z)r\,dz\,dr\,d\theta = \int_{\theta=0}^{2\pi}\int_{r=0}^{2}\int_{z=0}^{2}r^3 + 2zr\,dz\,dr\,d\theta = 2\pi\int_{r=0}^{2}2r^3 + 4r\,dr = 32\pi$

7. Use Stokes' theorem to solve exercise 4, section 4.1.
   Solution: Exercise 4 states: Find the path integral $\oint(3x+4y)dx + (2x+3y^2)dy$ over $x^2 + y^2 = 4$. Stokes' theorem states that the integral of the curl of a vector field integrated over an oriented surface (oriented by the right hand rule and the outward normal) is equal to the integral of the vector field over the boundary of the surface, moving over the boundary in the direction of the orientation, or $\int\int \nabla \times \mathbf{F} \cdot \mathbf{n}dS = \int \mathbf{F} \cdot d\mathbf{R}$. In this case we have $\mathbf{F} = (3x+4y)\mathbf{i} + (2x+3y^2)\mathbf{j}$, and $\nabla \times \mathbf{F} = -2\mathbf{k}$, so we calculate $\int\int -2\mathbf{k} \cdot d\mathbf{R} = \int_{\theta=0}^{2\pi}\int_{r=0}^{2}-2r\,dr\,d\theta = -8\pi$

9. Verify Stokes' theorem in the following special cases. Let C be the square in the xy plane with the equation $|x| + |y| = 1$. Let $\mathbf{F}$ be as follows:

   (a) $\mathbf{F} = x\mathbf{i}$
   (b) $\mathbf{F} = y\mathbf{i}$
   (c) $\mathbf{F} = -y\mathbf{i} + x\mathbf{j}$
   (d) $\mathbf{F} = \mathbf{i} + \mathbf{j}$
   (e) $\mathbf{F} = y^3\mathbf{i}$

   Solution: Referring to Figure 4.3, the sides of the square may be parameterized by

   $$(1-t)\mathbf{i} + t\mathbf{j}$$
   $$-t\mathbf{i} + (1-t)\mathbf{j}$$
   $$-(1-t)\mathbf{i} - t\mathbf{j}$$
   $$t\mathbf{i} - (1-t)\mathbf{j}$$

   for sides 1 through 4 respectively, with tangent vectors

   $$-\mathbf{i} + \mathbf{j}$$
   $$-\mathbf{i} - \mathbf{j}$$
   $$\mathbf{i} - \mathbf{j}$$
   $$\mathbf{i} + \mathbf{j}.$$

   a) We must compare $\int\int \nabla \times \mathbf{F} \cdot \mathbf{k}\,dy\,dx$ over the surface of the square and $\oint \mathbf{F} \cdot d\mathbf{R}$ over the sides. The curl of $\mathbf{F} = x\mathbf{i}$ is zero because derivatives are taken with respect to y and z, so the first integral is zero. In the second integral, the sum of the vector field $x\mathbf{i}$ dotted with the tangent vectors for sides one through four are easily seen to be zero. Therefore the integral over the boundary is also zero.

4.9. INTRODUCTION TO THE DIVERGENCE THEOREM AND STOKES' THEOREM

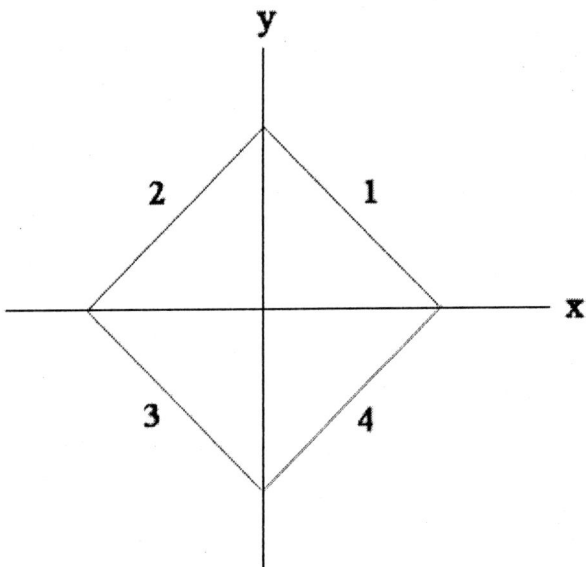

Figure 4.3: The square in problem 4.9.9.

b) In this case, the curl of the vector field is $-\mathbf{k}$, which, when dotted into $\mathbf{k}$ and integrated over the square of area $\sqrt{2}\sqrt{2} = 2$, gives -2. The vector field $\mathbf{F} = y\mathbf{i}$ is $t\mathbf{i}$, $(1-t)\mathbf{i}$, $-t\mathbf{i}$ and $-(1-t)\mathbf{i}$ for sides one through four, respectively, and the sum of the dot products with the tangent vectors $-\mathbf{i}+\mathbf{j}$, $-\mathbf{i}-\mathbf{j}$, $\mathbf{i}-\mathbf{j}$, and $\mathbf{i}+\mathbf{j}$ is $-t - (1-t) - t - (1-t) = -2$. The line integral $\oint_{t=0}^{t=1} -2dt = -2$ matches, so Stokes' theorem works again.

c) The curl of $\mathbf{F} = -y\mathbf{i} + x\mathbf{j}$ is $2\mathbf{k}$, which, when dotted into $\mathbf{k}$ and integrated over the square of area $\sqrt{2}\sqrt{2} = 2$, gives 4. On the four sides of the square, $\mathbf{F}$ is $-t\mathbf{i} + (1-t)\mathbf{j}$, $-(1-t)\mathbf{i} - t\mathbf{j}$, $t\mathbf{i} - (1-t)\mathbf{j}$ and $(1-t)\mathbf{i} + t\mathbf{j}$, respectively, and the sum of the dot products of these with $-\mathbf{i}+\mathbf{j}$, $-\mathbf{i}-\mathbf{j}$, $\mathbf{i}-\mathbf{j}$, and $\mathbf{i}+\mathbf{j}$ is $t + (1-t) + (1-t) + t + t + (1-t) + (1-t) + t = 4$, so the integral $\oint_{t=0}^{t=1} 4dt = 4$ and Stokes' theorem is confirmed in this special case.

d) It is easy to see that the curl of the constant vector field $\mathbf{F} = \mathbf{i} + \mathbf{j}$ is zero, so the integral of the curl over the surface is zero. Also, $\mathbf{F} \cdot \frac{d\mathbf{R}}{dt} = 0$ for each side of the square, so Stoke's theorem is confirmed for this special case.

e) Here the vector field $\mathbf{F} = y^3$ is $t^3\mathbf{i}$, $(1-t)^3\mathbf{i}$, $(-t)^3\mathbf{i}$ and $(-(1-t))^3\mathbf{i}$ for sides one through four respectively, which, when dotted with $-\mathbf{i}+\mathbf{j}$, $-\mathbf{i}-\mathbf{j}$, $\mathbf{i}-\mathbf{j}$, and $\mathbf{i}+\mathbf{j}$ yields $-t^3 - (1-t)^3 + (-t)^3 + (-(1-t))^3 = -2 + 6t - 6t^2$, which, when integrated from $t = 0$ to $t = 1$ is -1. The curl of $\mathbf{F} = y^3\mathbf{i}$ is $-2\mathbf{k}$, which when dotted into the outward normal to the square $\mathbf{k}$ is $-3y^2$. We can divide the integral into four parts, going clockwise beginning at the first quadrant: $\int_{x=0}^{x=1}\int_{y=0}^{y=1-x} -3y^2 dy dx$, $\int_{x=-1}^{x=0}\int_{y=0}^{y=1+x} -3y^2 dy dx$, $\int_{x=-1}^{x=0}\int_{y=-1-x}^{y=0} -3y^2 dy dx$ and $\int_{x=0}^{x=1}\int_{y=x-1}^{y=0} -3y^2 dy dx$. The inner integral is always y^3, so we must evaluate

$$-\left[\int_{x=0}^{x=1} y^3|_{y=0}^{y=1-x} dx + \int_{x=-1}^{x=0} y^3|_{y=0}^{y=1+x} dx + \int_{x=-1}^{x=0} y^3|_{y=-1-x}^{y=0} dx + \int_{x=0}^{x=1} y^3|_{y=x-1}^{y=0} dx\right] = -1.$$

11. Evaluate $\int\int_S (\nabla \times \mathbf{F}) \cdot d\mathbf{S}$, where $\mathbf{F} = y\mathbf{i} + (x - 2x^3z)\mathbf{j} + xy^3\mathbf{k}$ and S is the surface of a sphere $x^2 + y^2 + z^2 = a^2$ above the xy plane.
Solution: This calls for Stokes' theorem $\int\int_S (\nabla \times \mathbf{F}) \cdot d\mathbf{S} = \oint \mathbf{F} \cdot \frac{d\mathbf{R}}{dt} dt$. We will integrate around the circle $a\cos t\mathbf{i} + a\sin t\mathbf{j} + 0\mathbf{k}$, so $\mathbf{F} = a\sin t\mathbf{i} + a\cos t\mathbf{j} + a^4\cos t\sin^3 t\mathbf{k}$ and $\frac{d\mathbf{R}}{dt} = -a\sin t\mathbf{i} + a\cos t\mathbf{j} + 0\mathbf{k}$. We find $\mathbf{F} \cdot \frac{d\mathbf{R}}{dt} = a^2\cos^2 t - a^2\sin^2 t$, which can be seen to be 0 using symmetry arguments, that is, interpreting the integral as an area under a curve, and recall that the areas under $\cos^2 t$ and $\sin^2 t$ from 0 to 2π are the same. Alternately, integrate $\int_0^{2\pi} a^2\cos^2 t - a^2\sin^2 t dt = \cos t\sin t|_0^{2\pi} = 0$.

4.9. INTRODUCTION TO THE DIVERGENCE THEOREM AND STOKES' THEOREM

13. Evaluate $\int\int_S (\nabla \times \mathbf{F}) \cdot \mathbf{n} dS$, where $\mathbf{F} = 2y\mathbf{i} + (x - 2x^3 z)\mathbf{j} + xy^3\mathbf{k}$ and where S is the curved surface of the hemisphere $x^2 + y^2 + z^2 = 1$, $z \geq 0$.
    Solution: This is the same problem as number 11 stated differently and with the small changes $\mathbf{F} = 2y\mathbf{i} + (x - 2x^3 z)\mathbf{j} + xy^3\mathbf{k}$ and $x^2 + y^2 + z^2 = 1$, so we find $\int_0^{2\pi} \cos^2 t - 2\sin^2 t\, dt = \frac{1}{2}(3\cos t \sin t - t)|_0^{2\pi} = -\pi$.

15. Let $\mathbf{F}$ be the field $\mathbf{F} = ye^x\mathbf{i} + (x + e^x)\mathbf{j} + z^2\mathbf{k}$ and let C be the curve given by $\mathbf{R} = (1 + \cos t)\mathbf{i} + (1 + \sin t)\mathbf{j} + (1 - \sin t - \cos t)\mathbf{k}$ for $0 \leq t \leq 2\pi$. Find $\int_C \mathbf{F} \cdot d\mathbf{R}$ (*Hint:* Use Stokes' theorem observing that C is contained in a certain plane and that the projection of C on the xy plane is a circle.)
    Solution: We can see that this problem would look rather difficult if we substituted the parameterization of x, y and z along the path given into the vector field $\mathbf{F}$ and integrated along that path – there would be exponentials containing sines and cosines, so instead we integrate the curl of the flux of $\mathbf{F}$ over the simplest surface that is bounded by the closed path that we can find. The problem states that the parameterized path lies in a plane, so we can use the trick of projection of the path onto the xy plane using $\int\int_S \mathbf{F} \cdot d\mathbf{S} = \int\int \mathbf{F} \cdot \mathbf{n} \frac{dx\,dy}{|\cos\gamma|}$ (see pages 233 and 239). We find the normal to the plane in which $\mathbf{R}$ lies by finding two vectors tangent to the path and crossing them to get an outward normal. Tangent to the path is $\frac{d\mathbf{R}}{dt} = -\sin t \mathbf{i} + \cos t \mathbf{j} + (\sin t - \cos t)\mathbf{k}$, and two vectors tangent to the path, $\mathbf{v}_1 = \mathbf{j} - \mathbf{k}$, $\mathbf{v}_2 = -\mathbf{i} + \mathbf{k}$ are found by substituting 0 and $\pi/2$ into the expression for the tangent. Then an outward normal to the plane is $\mathbf{v}_1 \times \mathbf{v}_2 = \mathbf{i} + \mathbf{j} + \mathbf{k}$. The projection of $\mathbf{R}$ into the xy plane is $\mathbf{R}^* = (1 + \cos t)\mathbf{i} + (1 + \sin t)\mathbf{j}$, a circle of radius 1 centered at $(1,1)$. Thus, we integrate $\int\int \frac{(\nabla \times \mathbf{F}) \cdot \mathbf{n}}{|\cos\gamma|} dx\,dy = \int\int \frac{\mathbf{(k)} \cdot (\mathbf{i}+\mathbf{j}+\mathbf{k})\frac{1}{\sqrt{3}}}{|\frac{1}{\sqrt{3}}(\mathbf{k}) \cdot (\mathbf{i}+\mathbf{j}+\mathbf{k})|} dx\,dy = \pi$, which is just the area of the circle of radius 1.

17. Given $\phi(x, y, z) = xyz + 5$, find the surface integral of the normal component of $\mathbf{grad}\phi$ over $x^2 + y^2 + z^2 = 9$.
    Solution: Here we use the divergence theorem to find the integral of the divergence of $\mathbf{F} = \nabla\phi$ over the volume rather than the integral of the normal component of $\mathbf{F} = \nabla\phi$ over the surface. But $\mathbf{F} = yz\mathbf{i} + xz\mathbf{j} + xy\mathbf{k}$, and its divergence is 0, so the integral over the volume is 0.

19. Let $\mathbf{F} = \phi\nabla\phi$. Find the surface integral of the normal component of $\mathbf{F}$ over the surface of a sphere of radius 3 and center at the origin,

    (a) if $\phi = x + y + z$.
    (b) if $\phi = x^2 + y^2 + z^2$.

    Solution:

    (a) Because ϕ is harmonic, we can write $\int\int \phi\nabla\phi \cdot \mathbf{n}\,dS = \int\int\int \nabla \cdot (\phi\nabla\phi)dV = \int\int\int |\nabla\phi|^2 dV = 9\int\int\int dV = 9\frac{4}{3}\pi 3^2 = 108\pi$.

    (b) We can see that $\phi = x^2 + y^2 + z^2$ is not harmonic, so $\int\int \phi\nabla\phi \cdot \mathbf{n}\,dS = \int\int\int \nabla \cdot (\phi\nabla\phi)dV = \int\int\int \nabla \cdot (x^2 + y^2 + z^2)(2x\mathbf{i} + 2y\mathbf{j} + 2z\mathbf{k})dV$. Multiplying through and taking the divergence we get $10\int\int\int (x^2+y^2+z^2)dV$, the argument of which in spherical coordinates is $x^2+y^2+z^2 = r^2$, so we integrate $10\int_{\theta=0}^{\theta=2\pi}\int_{\phi=0}^{\phi=\pi}\int_{r=0}^{r=3} r^2 r^2 \sin\phi\,dr\,d\phi\,d\theta = 1944\pi$.

21. Despite the fact that the surface of exercise 6, section 4.7 is not closed, the divergence theorem can be used to reduce this problem to mental arithmetic. Show how to do this.
    Solution: Let W represent the missing top of the box and U the remainder of the surface. Then the divergence theorem tells us that $\int\int\int_V \nabla \cdot \mathbf{F}\,dV = \int\int_U \mathbf{F} \cdot d\mathbf{S} + \int\int_W \mathbf{F} \cdot d\mathbf{S}$. Because the divergence of $\mathbf{F}$ is zero, $\int\int_U \mathbf{F} \cdot d\mathbf{S} = -\int\int_W \mathbf{F} \cdot d\mathbf{S} = -\int\int (y\mathbf{i} + \mathbf{k}) \cdot \mathbf{k}\,dS = -\int_{x=0}^{x=1}\int_{y=0}^{y=1} 1\,dy\,dx = -1$.

23. One can compute the volume of a room by calculating the flux of the vector $\mathbf{R}$ through the walls. Show this.
    Solution: Let S be the surface bounding a volume V, where we are thinking of the surface as the walls of a room. By the divergence theorem, $\int\int \mathbf{F} \cdot d\mathbf{S} = \int\int\int \nabla \cdot \mathbf{F}\,dV$. For $\mathbf{F} = \mathbf{R} = x\mathbf{i} + y\mathbf{j} + z\mathbf{k}$, the divergence is $\nabla \cdot (x\mathbf{i} + y\mathbf{j} + z\mathbf{k}) = 3$ and $\int\int\int \nabla \cdot \mathbf{F}\,dV = 3V$, so the volume of a room is one third the flux of the vector field $x\mathbf{i} + y\mathbf{j} + z\mathbf{k}$ through the walls.

25. Given $\mathbf{F} = \frac{x\mathbf{i}+y\mathbf{j}+z\mathbf{k}}{x^2+y^2+z^2}$ find the surface integral of the normal component of $\mathbf{F}$ over the surface of the sphere $x^2 + y^2 + z^2 = 4$. Can you use the divergence theorem?

4.9. INTRODUCTION TO THE DIVERGENCE THEOREM AND STOKES' THEOREM

Solution: The divergence theorem $\int\int \mathbf{F}\cdot d\mathbf{S} = \int\int\int \nabla\cdot\mathbf{F}dV$ relates the flux through the surface to the divergence over the volume the surface encloses. In this case, $\nabla\times\mathbf{F} = \frac{3}{x^2+y^2+z^2} - 2\frac{(x^2+y^2+z^2)}{(x^2+y^2+z^2)^2} = \frac{3}{4} - \frac{1}{2} = \frac{1}{4}$. Therefore, the flux through the surface is $\frac{1}{4}$ the volume, or $\left(\frac{1}{4}\right)\left(\frac{4}{3}\right)\pi 2^3 = \frac{8}{3}\pi$.

27. Stokes' theorem provides an interesting interpretation of theorem 4.3, which identifies irrotational fields with conservative fields in simply connected domains. Show that if $\mathrm{curl}\mathbf{F} = 0$, then the line integral of $\mathbf{F}$ around any closed curve that bounds an oriented surface in the domain is zero. Where does simple connectedness come into play?

Solution: By Stokes' theorem, $\int\int \nabla\times\mathbf{F}\cdot d\mathbf{S} = \int \mathbf{F}\cdot d\mathbf{R}$ and if $\nabla\times\mathbf{F} = 0$, then the integral over the boundary of any closed simply connected surface in the domain will also be zero. Simple connectedness comes into play because if there is a hole space then for some surfaces there will be a hole in the surface, in which cases it will have two boundaries.

29. The abstract concept of a gooney sphere comes from the shape of a gooney egg. A gooney bird is born with a pointed head and a prominent stubby tail; therefore the shape of the egg is ellipsoidal with pointed ends. Surface integrals over gooney spheres are difficult to compute; tables of gooney functions are needed, but these were tabulated during the war and are still classified top secret. All that is known is that a gooney sphere of minimal diameter $d = 1$ has volume approximately 0.7.

(a) Find the surface integral of the normal component of $\mathbf{F} = x\mathbf{i} + y\mathbf{j} + z\mathbf{k}$ over the surface of a gooney sphere with center at the origin and minimal diameter $d = 2$, making any assumptions you deem reasonable.

(b) Would your answer be the same if the gooney sphere had center at $(2, 7, -3)$?

Solution:

(a) Using the divergence theorem, we know the flux through the surface of $\mathbf{F} = x\mathbf{i} + y\mathbf{j} + z\mathbf{k}$ is equal to the divergence, which is 3 over the volume. Assuming the shape of eggs is constant as the diameter changes, and that the eggs' volumes scale with the cube of the minimal diameter, so that $Kd^3 = V$, and with $K(1)^3 \approx 0.7$, $K \approx 0.7$. Therefore, the volume of a gooney egg of minimal diameter 2 has volume $2^3 \cdot 0.7 = 5.6$. The flux through the shell is three times this, or about 16.8.

(b) The answer would be the same no matter the origin, because the divergence of the field $\mathbf{F}$ is independent of position.

31. Surface and volume integrals of vector valued functions are defined as for numerically valued (scalar valued) functions. Alternatively they can be defined by simply integrating separately the x, y and z components (which are numerical). Show formally that $\int\int\int_D \nabla\phi dV = \int\int_S \phi\mathbf{n}dS$ by applying the divergence theorem to $\mathbf{F} = \phi\mathbf{C}$, where $\mathbf{C}$ is a constant vector field.

Solution: Integrate the divergence of $\mathbf{F} = \phi\mathbf{C}$ over the volume $\int\int\int \nabla\cdot(\phi\mathbf{C})dV = \int\int\int \phi\nabla\cdot\mathbf{C} + \mathbf{C}\cdot\nabla\phi dV = \int\int\int \mathbf{C}\cdot\nabla\phi dV = \mathbf{C}\cdot\int\int\int \nabla\phi dV$. Also, $\int\int \phi\mathbf{C}\cdot \mathbf{n}dS = \mathbf{C}\cdot\int\int \phi\mathbf{n}dS$, so $\mathbf{C}\cdot\int\int\int \nabla\phi dV = \mathbf{C}\cdot\int\int \phi\mathbf{n}dS$. Since this is true for any constant vector field $\mathbf{C}$, $\int\int\int \nabla\phi dV = \int\int \phi\mathbf{n}dS$.

33. Give a vector interpretation of each of the following.

(a) $\lim\limits_{V\to 0} \frac{1}{V} \int\int_S \mathbf{n}fdS$

(b) $\lim\limits_{V\to 0} \frac{1}{V} \int\int_S \mathbf{n}\times\mathbf{F}dS$

Solution:

(a) This is the gradient of f. By problem 31 in this section, $\int\int\int_D \nabla f dV = \int\int_S \mathbf{n}f dS$. In the x direction, this is $\int\int\int_D \nabla f\cdot\mathbf{i}dV = \int\int_S \mathbf{n}f\cdot\mathbf{i}dS$. By the mean value theorem for integrals, $\frac{1}{V}\int\int\int_D \nabla f\cdot\mathbf{i}dV = \nabla f\cdot\mathbf{i}|_{p^*}$ for some point $p^* = (x^*, y^*, z^*)$ in V, so $\nabla f\cdot\mathbf{i}|_{p^*} = \frac{1}{V}\int\int_S \mathbf{n}f dS$. Now take the limit as the volume V goes to zero, and we have the result for the component of the gradient in the x direction. Making the same argument for the y and z directions, we obtain the result.

4.10. INTRODUCTION TO THE TRANSPORT THEOREMS

(b) This is the curl of **F**. By problem 32 in this section, $\iiint_D \nabla \times \mathbf{F}\, dV = \iint_S \mathbf{n} \times \mathbf{F}\, dS$. In the x direction, this is $\iiint_D \nabla \times \mathbf{F} \cdot \mathbf{i}\, dV = \iint_S \mathbf{n} \times \mathbf{F} \cdot \mathbf{i}\, dS$. By the mean value theorem for integrals, $\frac{1}{V}\iiint_D \nabla \times \mathbf{F} \cdot \mathbf{i}\, dV = \nabla \times \mathbf{F} \cdot \mathbf{i}|_{p^*}$ for some point $p^* = (x^*, y^*, z^*)$ in V, so $\nabla \times \mathbf{F} \cdot \mathbf{i}|_{p^*} = \frac{1}{V}\iint_S \mathbf{n} \times \mathbf{F}\, dS$. Now take the limit as the volume V goes to zero, and we have the result for the component of the curl in the x direction. Making the same argument for the y and z directions, we obtain the result.

35. Let **J** denote the electric current density (a vector in the direction of the current (a vector in the direction of the current, with magnitude in units of current per area) and **B** denote the magnetic field intensity. One of Maxwell's laws of electromagnetism states that, in the absence of a time varying electric field, $\mathbf{curl B} = \mu_0 \mathbf{J}$ where μ_0 is a constant. Use Stokes' theorem to derive $\int_C \mathbf{B} \cdot d\mathbf{R} = \mu_0 I$. In words: the line integral of the tangential component of the magnetic field intensity, around a closed loop, is proportional to the current I passing across any surface bounded by the loop.
Solution: By Stokes' theorem $\iint \nabla \times \mathbf{B} \cdot d\mathbf{S} = \int \mathbf{B} \cdot d\mathbf{R}$, so $\iint \mu_0 \mathbf{J} \cdot d\mathbf{S} = \int \mu_0 \mathbf{B} \cdot d\mathbf{R}$. But $\iint \mathbf{J} \cdot d\mathbf{S}$ is just the current density vector with units of current per unit area, integrated over an area, so $\iint \mathbf{J} \cdot d\mathbf{S} = I$, the current through the surface bounded by the loop. Therefore $\int \mu_0 \mathbf{B} \cdot d\mathbf{R} = \mu_0 I$.

4.10 Introduction to the Transport Theorems

1. Let S_t be a uniformly expanding hemisphere described by $x^2 + y^2 + z^2 = (\nu t)^2$, $(z \geq 0)$ and let **F** be the vector field $\mathbf{F}(\mathbf{R}, t) = \mathbf{R}t$. Verify the flux of transport theorem in this case.
Solution: The flux through the moving surface is $\Phi = \iint_{S_t} \mathbf{F}(\mathbf{R}, t) \cdot d\mathbf{S}$, and the flux transport theorem states

$$\frac{d\Phi}{dt} = \iint_{S_t} \left(\frac{\partial \mathbf{F}}{\partial t} + (\nabla \cdot \mathbf{F})\mathbf{v}\right) \cdot d\mathbf{S} + \oint_{C_t} \mathbf{F} \times \mathbf{v} \cdot d\mathbf{R} = \iint_{S_t} \left(\frac{\partial \mathbf{F}}{\partial t} + (\nabla \cdot \mathbf{F})\mathbf{v} + \nabla \times (\mathbf{F} \times \mathbf{v})\right) \cdot d\mathbf{S}.$$

In this exercise, the region is expanding radially at a rate of ν and the vector field is a radial field whose strength depends on the radius and time as $\mathbf{F} = rt\mathbf{e}_r$ in spherical coordinates. Because the field and the outward normal vector **v** point in the same direction, the vector product $\mathbf{F} \times \mathbf{v}$ will be zero. The position vector describing the surface in spherical coordinates is $\mathbf{R} = \nu t \mathbf{e}_r$, and the velocity vector is $\mathbf{v} = \nu \mathbf{e}_r$. The vector field is $\mathbf{F} = \mathbf{F}(\mathbf{R}, t) = rt\mathbf{e}_r = \nu t^2 \mathbf{e}_r$, and the outward unit normal to the surface is $\mathbf{e}_r$. Therefore, we have

$$\Phi = \iint_{S_t} \mathbf{F} \cdot \mathbf{n}\, dS = \iint_{S_t} \nu t^2 \mathbf{e}_r \cdot \mathbf{e}_r r^2 \sin\phi\, d\phi\, d\theta = \int_{\theta=0}^{2\pi}\int_{\phi=0}^{\pi/2} \nu t^2 \nu^2 t^2 \sin\phi\, d\phi\, d\theta = 2\pi \nu^3 t^4,$$

whose derivative is $8\pi\nu^3 t^3$. Computing the same thing from the transport theorem,

$$\frac{\partial \Phi}{\partial t} = \iint_{S_t}\left(\frac{\partial \mathbf{F}}{\partial t} + (\nabla \cdot \mathbf{F})\mathbf{v} + \nabla \times (\mathbf{F} \times \mathbf{v})\right) \cdot \mathbf{n}\, dS = \int_{\theta=0}^{2\pi}\int_{\phi=0}^{\pi/2} (r\mathbf{e}_r + (3t)\nu\mathbf{e}_r) \cdot \mathbf{e}_r r^2 \sin\phi\, d\phi\, d\theta$$

$$= \int_{\theta=0}^{2\pi}\int_{\phi=0}^{\pi/2} 4\nu^3 t^3 \sin\phi\, d\phi\, d\theta = 8\pi\nu^3 t^3.$$

3. Suppose the square $0 \leq x \leq 1$, $0 \leq y \leq 1$ is rotated about the x axis at a constant angular velocity. Verify the flux transport theorem with the uniform vector field $\mathbf{F}(\mathbf{R}, t) = \mathbf{k}$.
Solution: We will use the following form of the theorem:

$$\frac{d\Phi}{dt} = \frac{d}{dt}\iint_{S_t} \mathbf{F}(\mathbf{R}, t) \cdot \mathbf{n}\, dS = \iint_{S_t}\left(\frac{\partial \mathbf{F}}{\partial t} + (\nabla \cdot \mathbf{F})\mathbf{v}\right) \cdot \mathbf{n}\, dS + \oint_{C_t} \mathbf{F} \times \mathbf{v} \cdot d\mathbf{R},$$

and rotate the surface counterclockwise about the x axis. With a uniform constant vector field $\mathbf{F} = \mathbf{k}$ pointing in the z direction, we expect the flux through the rotating surface to be periodic,

4.10. INTRODUCTION TO THE TRANSPORT THEOREMS

indeed, the unit normal to the rotating surface is $-\sin kt\mathbf{j} + \cos kt\mathbf{k}$, for constant k so $\int\int_{S_t} \mathbf{k} \cdot (-\sin kt\mathbf{j} + \cos kt\mathbf{k})dS = \cos kt \int\int dS = \cos kt$ because the area of the square is 1. Therefore, $\frac{\partial \Phi}{\partial t} = -k\sin kt$. On the right hand side, $\mathbf{F}$ is constant so its time derivative will be zero and the first term disappears. Also, the second term will disappear because the divergence of a constant vector field is zero. Finally, around the perimeter of the square, the portion in the x axis has velocity zero, and the sides parallel to the yz plane have identical velocities but opposite orientations as we traverse the boundary, and as the vector field is constant in space and time, their contributions to the line integral vanish. Finally, the side of the square parallel to the x axis a distance 1 away from that axis has velocity $-k\sin kt\mathbf{j} + k\cos kt\mathbf{k}$, and $\frac{d\mathbf{R}}{ds}$ along this side is $-\mathbf{i}$. Putting this all together, we find $\oint_{C_t} (\mathbf{F} \times \mathbf{v}) \cdot \frac{d\mathbf{R}}{ds} ds = \oint_{C_t} (k\sin kt\mathbf{i}) \cdot (-\mathbf{i})ds = -k\sin kt$, agreeing with the integral on the left hand side.

5. Verify Reynold's theorem for a unit cube with edges parallel to the axes, sliding in the x direction at a constant velocity and with $\rho(\mathbf{R}, t) = xy$.
   Solution: We assume the author means $\eta(\mathbf{R}, t)$ rather than $\rho(\mathbf{R}, t)$. From a physical standpoint, the nature of the scalar field is irrelevant here; it could be temperature, pressure, mass density, momentum density, etc. Reynold's transport theorem has several forms:

   (a) $\frac{d}{dt}\int\int\int_{V_t} \eta(\mathbf{R}, t)dV = \int\int\int_{V_t} \left(\frac{d\eta}{dt} + \eta \nabla \cdot \mathbf{v}\right) dV$

   (b) $\frac{d}{dt}\int\int\int_{V_t} \eta(\mathbf{R}, t)dV = \int\int\int_{V_t} \left(\frac{d\eta}{dt} + \nabla \cdot (\eta\mathbf{v})\right) dV$

   (c) $\frac{d}{dt}\int\int\int_{V_t} \eta(\mathbf{R}, t)dV = \int\int\int_{V_t} \frac{d\eta}{dt} dV + \int\int_{S_t} \eta\mathbf{v} \cdot d\mathbf{S}$

   For a moving volume sliding in the x direction, xy can be written $(x_0 + kt)y$, and $\mathbf{v} = kt\mathbf{i}$. The integral of the left hand side is $\int\int\int_{V_t}(x_0 + kt)y dx\, dy\, dz = \frac{1}{2}y^2 x_0 z + \frac{1}{2}y^2 xktz$, with derivative $\frac{1}{2}kxy^2 z$. The right hand side is $\int\int\int \frac{\partial}{\partial t}(x_0 + kt)y + (x_0 + kt)y\nabla \cdot i dV = \int\int\int dV = \int\int\int ky + 0 dx\, dy\, dz = \frac{1}{2}kxy^2 z$.

7. In a so-called *perfect fluid*, the *pressure* $p(\mathbf{R}, t)$ exerts a force per unit area on a surface given by $-p\mathbf{n}$, where $\mathbf{n}$ is the unit surface normal. Gravity exerts a force per unit mass on the fluid given by the constant vector $\mathbf{g}$ (the gravitational acceleration).

   (a) Show that Newton's second law, applied to the volume of a perfect fluid, is expressed

   $$\frac{d}{dt}\int\int\int_V \mu\mathbf{v} dV = \int\int\int_V \mu\mathbf{g} dV - \int\int_S p\mathbf{n} \cdot d\mathbf{S}$$

   where S is the surface boundary of V.

   (b) Modify the analysis of section 3.3 to rewrite the pressure force in terms of the gradient; then use Cauchy's expression for the momentum term and derive

   $$\int\int\int_V \mu \frac{d}{dt}\mathbf{v} dV = \int\int\int_V \{\mu\mathbf{g} - \nabla p\} dV.$$

   (c) Argue that (4.59) implies the equation of motion $\mu \frac{d}{dt}\mathbf{v} = \mu\mathbf{g} - \nabla p$.

   (d) For steady flow all the Eulerian time derivatives ($\frac{\partial}{\partial t}$) are zero. Use exercise 14, section 3.8, to derive the equation for steady flow $\mu\mathbf{v} \cdot \nabla\mathbf{v} = \mu\mathbf{g} - \nabla p$.

   Solution:

   (a) By Newton's second law, $F = ma = m\mathbf{g} - p\mathbf{n}$. The total force over a volume is the gravitational force over the volume plus the pressure force over the surface. Writing $\mathbf{a} = \frac{d}{dt}\mathbf{v}$ and noting that mass equals density μ times volume, we integrate $m\frac{d}{dt}\mathbf{v}$ and $m\mathbf{g}$ over the volume and $-p\mathbf{n}$ over the surface of the volume and add them to get $\int\int\int_V \mu \frac{d}{dt}\mathbf{v} dV = \int\int\int_V \mu\mathbf{g} dV - \int\int_S p\mathbf{n} dS$. By exercise 6 of this section, $\frac{d}{dt}\int\int\int_V \mu\mathbf{F}dV = \int\int\int_V \mu\frac{d\mathbf{F}}{dt} dV$ (note that the book has $d\mathbf{V}$ as a vector, which is a misprint). Therefore $\frac{d}{dt}\int\int\int_V \mu\mathbf{v}dV = \int\int\int_V \mu\mathbf{g} dV - \int\int_S p\mathbf{n} \cdot d\mathbf{S}$.

4.10. INTRODUCTION TO THE TRANSPORT THEOREMS

(b) Consider the pressure force on the walls of a small cube with side lengths Δx, Δy and Δz oriented so that the sides are aligned with the xy, yz and xz planes. The total pressure over the faces parallel to the yz plane is

$$[p(x + \Delta x, y, z)\Delta y \Delta z \mathbf{i} + p(x, y, z)\Delta y \Delta z(-\mathbf{i})]\,\mathbf{i},$$

and likewise for the sides parallel to the xy and zx planes,

$$[p(x, y + \Delta y, z)\Delta x \Delta z \mathbf{j} + p(x, y, z)\Delta x \Delta z(-\mathbf{j})]\,\mathbf{j}$$

and

$$[p(x, y, z + \Delta z)\Delta y \Delta x \mathbf{k} + p(x, y, z)\Delta y \Delta x(-\mathbf{k})]\,\mathbf{k}.$$

This corresponds to the integral $\int\int_S p\mathbf{n}\cdot d\mathbf{S}$, and in the limit as the area elements approach zero, becomes exact.

Writing the pressure differences as

$$\frac{p(x+\Delta x, y, z) - p(x,y,z)}{\Delta x}\Delta x \Delta y \Delta z,$$

$$\frac{p(x, y+\Delta y, z) - p(x,y,z)}{\Delta y}\Delta x \Delta y \Delta z$$

and

$$\frac{p(x, y, z+\Delta z) - p(x,y,z)}{\Delta z}\Delta x \Delta y \Delta z$$

and taking the limit as Δx, Δy and Δz go to zero, we see we have $\frac{\partial p}{\partial x}dV$, $\frac{\partial p}{\partial y}dV$, and $\frac{\partial p}{\partial z}dV$, or

$$\left[\frac{\partial p}{\partial x}\mathbf{i} + \frac{\partial p}{\partial y}\mathbf{j} + \frac{\partial p}{\partial z}\mathbf{k}\right]\cdot d\mathbf{V} = \nabla p \cdot d\mathbf{V},$$

which, evaluated over a volume is $\int\int\int_V \nabla p \cdot d\mathbf{V}$. Cauchy's expression for the rate of change of momentum is $\frac{d}{dt}\int\int\int_V \mu \mathbf{v}\, dV = \int\int\int_V \mu \frac{d\mathbf{v}}{dt}\, dV$, so putting this all together, we get

$$\int\int\int_V \mu \frac{d}{dt}\mathbf{v}\, dV = \int\int\int_V \{\mu \mathbf{g} - \nabla p\}\, dV.$$

(c) Because the integral is over an arbitrary volume and for the volume at any location, the arguments are equal, so $\mu \frac{d}{dt}\mathbf{v} = \mu g - \nabla p$.

(d) By equation problem 14 in section 3.8, $\frac{d\mathbf{v}}{dt} = \frac{\partial \mathbf{v}}{\partial t} + \mathbf{v}\cdot\nabla\mathbf{v}$, but because $\frac{\partial \mathbf{v}}{\partial t} = 0$, $\frac{d\mathbf{v}}{dt} = \mathbf{v}\cdot\nabla\mathbf{v}$ so $\mu\mathbf{v}\cdot\nabla\mathbf{v} = \mu\mathbf{g} - \nabla p$.

Chapter 5

Advanced Topics

5.1 The Divergence Theorem

1. In what point in the proof of the divergence theorem did we make use of the requirement that the partial derivative $\frac{\partial F_3}{\partial z}$ be a continuous function of z.
   Solution: When the fundamental theorem of calculus was used, which requires the function be continuous.

3. Show, by a diagram similar to that of figure 5.1, that the volume integral of a function, taken over D, can be obtained by first integrating with respect to z and then integrating over the projection of S on the xy plane.
   Solution:

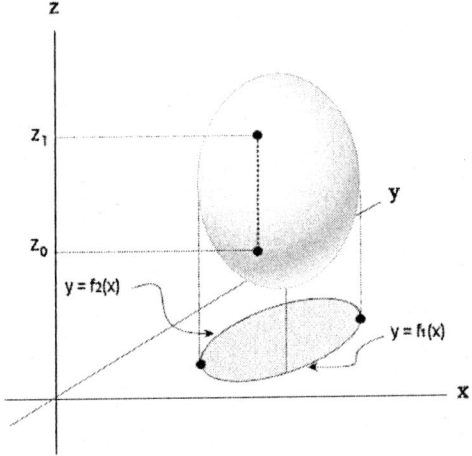

Figure 5.1: A convex blob.

5. Where in your "proof" (exercise 4) did you make unconscious use of the fact that the points on S with normals parallel to the xy plane have a projection on the xy plane of zero area? (*Hint:* look again at the definition of the area of a surface (section 4.6). What is $\cos \gamma$ for such points?)
   Solution: Referring to the equation $\int \int \int \frac{\partial F_3}{\partial z} dz\, dx\, dy = \int \int (F_3 \mathbf{k}) \cdot \mathbf{n}\, dS$, when integrating $\frac{\partial F_3}{\partial z}$ over a volume we integrate from the upper surface to the lower, which are at a distance of 0 from each other at the boundary of the surface being integrated over on the right hand side of the equation. Thus they contribute nothing to the volume integral on the left hand side. It is easy to

5.2. GREEN'S FORMULAS: LAPLACE'S AND POISSON'S EQUATION

see that $\cos\gamma = 0$ for these points (also $\mathbf{k}\cdot\mathbf{n} = 0$ there), so there is no "double counting" when we integrate up to this boundary on both the upper and the lower surface on the right hand side of this equation, because the integral is zero there also.

7. What is the flux output from an ellipsoid of volume v if $\mathbf{F} = 3x\mathbf{i} + y\mathbf{j} + z\mathbf{k}$?
   Solution: Because the divergence of $\mathbf{F}$ is 5, the integral of the flux over the surface is $5v$ by the divergence theorem.

9. (a) Describe the oriented surface enclosing the region $1 \leq x^2 + y^2 + z^2 \leq 4$, assuming the usual convention concerning the orientation of a closed surface. (In section 4.6 it was mentioned that if a surface encloses a region of space, the unit normal points away from the enclosed region; in this problem, the surface has two disconnected parts.)
   (b) How would you compute the surface integral of the normal component of a vector field $\mathbf{F}$ over this surface?
   (c) If div $\mathbf{F} = 0$ except perhaps at the origin, what can you say about the flux $\int\int \mathbf{F}\cdot\mathbf{n}dS$ over the two parts forming this surface, taking $\mathbf{n}$ to be the unit normal outward from the origin in each case?
   (d) Determine whether your answer to (c) would be any different if the region were that between the sphere $x^2 + y^2 + z^2 = 1$ and the ellipsoid $\frac{x^2}{4} + \frac{y^2}{9} + \frac{z^2}{16} = 1$.
   (e) Compute the surface integral of the normal component of $\mathbf{F} = \frac{x\mathbf{i}+y\mathbf{j}+z\mathbf{k}}{(x^2+y^2+x^2)^{3/2}}$ over the ellipsoid $\frac{x^2}{4} + \frac{y^2}{9} + \frac{z^2}{16} = 1$.
   Solution:

   (a) The surface consists of two parts, an outer sphere of radius 2 with the outward normal $\mathbf{n}$ pointing away from the origin and an inner sphere of radius 1 with outward normal $\mathbf{n}$ pointing toward the origin. Remember that an outward normal to a boundary enclosing a volume points away from the content of the enclosed region.

   (b) Let O be the outer surface and I be the inner surface. Then the surface integral is the sum of two integrals over the two surfaces $\int\int_O \mathbf{F}\cdot\mathbf{n}dS + \int\int_I \mathbf{F}\cdot\mathbf{n}dS$.

   (c) The outward fluxes must be equal and opposite so that their sum is equal to the integral of zero divergence over the volume.

   (d) No. If the divergence over the volume is zero, the sums of the fluxes over the surfaces must balance, no matter the shape.

   (e) Because the divergence is 0 except at the origin, the integral of the flux over the surface of the sphere and over the surface of the ellipsoid match, so $\int\int_{\text{ellipsoid}} \mathbf{F}\cdot\mathbf{n}dS = \int\int_{\text{sphere}} \mathbf{F}\cdot\mathbf{n}dS = \int\int_{\text{sphere}} \mathbf{R}/R^3\cdot\mathbf{R}/RdS = \int\int_{\text{sphere}} dS = 4\pi$.

5.2 Green's Formulas: Laplace's and Poisson's Equation

1. Evaluate $\int\int_S \left[\phi\nabla\left(\frac{1}{R}\right) - \frac{1}{R}\nabla\phi\right]\cdot d\mathbf{S}$ over the surface of the sphere $(x-3)^2 + y^2 + z^2 = 25$, where $\phi = xyz + 5$.
   Solution: We can apply the third Green Formula. $\pi\phi(0,0,0) = \frac{-1}{4\pi}\int\int\int_D \frac{\nabla^2\phi}{R}dV + \frac{1}{4\pi}\int\int_S\left[\frac{\nabla\phi}{R} - \phi\nabla\left(\frac{1}{R}\right)\right]\cdot\mathbf{n}dS$. To begin, we need to find $\nabla^2\phi$.
   $\nabla^2\phi = \frac{\partial^2\phi}{\partial x}\mathbf{i} + \frac{\partial^2\phi}{\partial y}\mathbf{j} + \frac{\partial^2\phi}{\partial z}\mathbf{k} = 0\mathbf{i} + 0\mathbf{j} + 0\mathbf{k}$
   Therefore, $\frac{-1}{4\pi}\int\int\int_D \frac{\nabla^2\phi}{R}dV = 0$. So, the original equation reduces to $\pi\phi(0,0,0) = \frac{1}{4\pi}\int\int_S\left[\frac{\nabla\phi}{R} - \phi\nabla\left(\frac{1}{R}\right)\right]\cdot\mathbf{n}dS$. Finally, we can solve for $\int\int_S\left[\frac{\nabla\phi}{R} - \phi\nabla\left(\frac{1}{R}\right)\right]\cdot\mathbf{n}dS = 4\pi\phi(0,0,0)$. And, $4\pi\phi(0,0,0) = 4\pi(0\cdot 0\cdot 0 + 5) = 20\pi$. So, $\int\int_S\left[\phi\nabla\left(\frac{1}{R}\right) - \frac{1}{R}\nabla\phi\right]\cdot d\mathbf{S} = 20\pi$.

3. Show that equation 5.16 follows by interpreting R as $|\mathbf{R}-\mathbf{R}'|$ in lemmas 5.1 through 5.3.
   Solution: Equation (5.16) states

5.2. GREEN'S FORMULAS: LAPLACE'S AND POISSON'S EQUATION

$\phi(\mathbf{R}) = \frac{-1}{4\pi} \iiint_D \frac{\nabla'^2 \phi(\mathbf{R}')}{|\mathbf{R}-\mathbf{R}'|} dV' + \frac{1}{4\pi} \iint_S \left[\frac{\nabla' \phi(\mathbf{R}')}{|\mathbf{R}-\mathbf{R}'|} - \phi(\mathbf{R}')\nabla'\left(\frac{1}{|\mathbf{R}-\mathbf{R}'|}\right) \right] \cdot \mathbf{n}' dS'$. Rewriting the proofs of each of Lemmas 5.1 through 5.3 in the context of a sphere located at $\mathbf{R}$, instead of at the origin, we have, taking care to note that $\mathbf{R} = a\mathbf{i} + b\mathbf{j} + c\mathbf{k}$ is now a *constant* vector while $\mathbf{R}' = x'\mathbf{i} + y'\mathbf{j} + z'\mathbf{k}$ varies,

(a) Let S be a sphere of radius $b = |\mathbf{R} - \mathbf{R}'|$ centered at the origin, and let $f(\mathbf{R} - \mathbf{R}')$ be a continuous scalar field. Then $\lim_{\mathbf{R}' \to \mathbf{R}} \iint \frac{f}{|\mathbf{R}-\mathbf{R}'|} dS = 0$. *Proof:* The integral $\iiint f dS$ equals the average value of f on the sphere times the area $4\pi|\mathbf{R}-\mathbf{R}'|^2$. In the limit, we have $\lim_{\mathbf{R}' \to \mathbf{R}} \frac{4\pi|\mathbf{R}-\mathbf{R}'|^2}{|\mathbf{R}-\mathbf{R}'|} = \lim_{\mathbf{R}' \to \mathbf{R}} 4\pi|\mathbf{R}-\mathbf{R}'| = 0$.

(b) Let S and f be as in Lemma 5.1. Then $\lim_{\mathbf{R}' \to \mathbf{R}} \iint_S f \nabla\left(\frac{1}{|\mathbf{R}-\mathbf{R}'|}\right) \cdot \mathbf{n} dS = -4\pi f(a,b,c)$. *Proof:* The outward normal on the sphere is $\mathbf{n} = \frac{\mathbf{R}'-\mathbf{R}}{|\mathbf{R}'-\mathbf{R}|}$ and
$\nabla'(1/|\mathbf{R}' - \mathbf{R}|) = \nabla'\left((x'-a)^2 + (y'-b)^2 + (z'-c)^2\right)^{-1/2} = -\frac{\mathbf{R}'-\mathbf{R}}{(|\mathbf{R}'-\mathbf{R}|)^3}$, so

$$\nabla \frac{1}{|\mathbf{R}'-\mathbf{R}|} \cdot \mathbf{n} = -\frac{|\mathbf{R}'-\mathbf{R}|^2}{|\mathbf{R}'-\mathbf{R}|^4} = -\frac{1}{|\mathbf{R}'-\mathbf{R}|^2}$$

on the sphere. The integral $\iint f dS$ is the average value of f on the sphere multiplied by the area $4\pi|\mathbf{R}'-\mathbf{R}|^2$, so as $|\mathbf{R}'-\mathbf{R}|$ goes to zero, the limit of the integral approaches $f(a,b,c)$.

(c) The laplacian of $\frac{1}{R}$ is zero, except at the origin: $\nabla^2\left(\frac{1}{R}\right) = 0$ for $(R \neq 0)$. *Proof:* We have already calculated $\nabla'(1/|\mathbf{R}' - \mathbf{R}|) = -\frac{\mathbf{R}'-\mathbf{R}}{(|\mathbf{R}'-\mathbf{R}|)^3}$, so we calculate
$\nabla'^2(1/|\mathbf{R}' - \mathbf{R}|) = -\nabla' \cdot \left[\frac{\mathbf{R}'-\mathbf{R}}{(|\mathbf{R}'-\mathbf{R}|)^3}\right] = -\nabla' \cdot (\mathbf{R}' - \mathbf{R})\left((x'-a)^2 + (y'-b)^2 + (z'-c)^2\right)^{-3/2}$
$- (\mathbf{R}' - \mathbf{R}) \cdot \nabla \left((x'-a)^2 + (y'-b)^2 + (z'-c)^2\right)^{-3/2}$. Expanding this, we find $\nabla'^2(1/|\mathbf{R}' - \mathbf{R}|) = -\frac{3}{|\mathbf{R}'-\mathbf{R}|^3} - (\mathbf{R}' - \mathbf{R}) \cdot \left(-3\frac{\mathbf{R}'-\mathbf{R}}{|\mathbf{R}'-\mathbf{R}|^5}\right) = 0$.

Now we apply the second Green formula with $\Phi = \frac{1}{|\mathbf{R}'-\mathbf{R}|}$ to a volume D enclosing the point (a,b,c), excluding a small sphere D' centered at (a,b,c): $\iiint_{D-D'} \left[\phi \nabla'^2\left(\frac{1}{|\mathbf{R}'-\mathbf{R}|}\right) - \frac{1}{|\mathbf{R}'-\mathbf{R}|}\nabla'^2 \phi\right] dV = \iint_S \left[\phi \nabla'\left(\frac{1}{|\mathbf{R}'-\mathbf{R}|}\right) - \frac{1}{|\mathbf{R}'-\mathbf{R}|}\nabla' \phi\right] \cdot \mathbf{n} dS + \iint_{S'} \left[\phi \nabla'\left(\frac{1}{|\mathbf{R}'-\mathbf{R}|}\right) - \frac{1}{|\mathbf{R}'-\mathbf{R}|}\nabla' \phi\right] \cdot \mathbf{n} dS$ where as before S is the boundary of D and S' is the boundary of D'. In the limit as the radius of D' goes to zero, and using the modified lemmas, we obtain equation (5.16) $\phi(\mathbf{R}) = \frac{-1}{4\pi} \iiint_D \frac{\nabla'^2 \phi(\mathbf{R}')}{|\mathbf{R}-\mathbf{R}'|} dV' + \frac{1}{4\pi} \iint_S \left[\frac{\nabla' \phi(\mathbf{R}')}{|\mathbf{R}-\mathbf{R}'|} - \phi(\mathbf{R}')\nabla'\left(\frac{1}{|\mathbf{R}-\mathbf{R}'|}\right)\right] \cdot \mathbf{n}' dS'$.

5. Prove that the solutions to Poisson's and Laplace's equations are uniquely determined in D by the values of their normal derivatives on S and the value of the solution at one (any) point of D.
Solution: Poisson's equation is $\nabla^2 \phi = \rho$, and Laplace's equation is $\nabla^2 \phi = 0$. Let $F = \phi = \psi$ be a solution to Laplace's equation and substitute into the first Green formula

$$\iiint_D \phi \nabla^2 \psi \, dV = \iint_S \phi \nabla \psi \cdot \mathbf{n} - \iiint_D \nabla \phi \cdot \nabla \psi \, dV$$

to get $\iiint_D F \nabla^2 F \, dV = \iint_S F \nabla F \cdot \mathbf{n} - \iiint_D |\nabla F| dV$. If the normal derivative on S is zero, and because F solves Laplace's equation, we have $\iiint_D |\nabla F| dV = 0$, which requires that $\nabla F = 0$, or that F be constant. Now let f and g both solve Poisson's equation and be such that their normal derivatives (directional derivatives in the direction of the surface normal $\nabla f \cdot \mathbf{n}$) agree on the boundary. Then $\phi = \psi = f - g$ solves Laplace's equation, and substitution into the first Green's formula gives us $\iiint_D |\nabla (f-g)| dV = 0$, or that f and g differ by a constant. But their normal derivatives agree on the boundary, so they agree everywhere. Because all functions identical on the boundary differ in the interior of the region D by a constant, all that we need is a single value from the interior to determine this constant and the function is uniquely determined.

7. The definition of the Green's function in the text actually describes the "Green's function of the first kind." The "Green's function of the second kind" takes the same form $1/R + \gamma(\mathbf{R})$ with γ harmonic throughout D, but with the boundary values of $\nabla G \cdot \mathbf{n}$ (rather than those of G itself) chosen so as to simplify the third Green formula [eq. (5.15)].

5.3. THE FUNDAMENTAL THEOREM OF VECTOR ANALYSIS

(a) What would result if we set $\psi = G$ in the second Green formula [eq. (5.10)] for a G making $\nabla G \cdot \mathbf{n} = 0$ on S?

(b) Derive the identity $\int\int_S \nabla G \cdot \mathbf{n} dS = -4\pi$ and show why this makes the G described in part (a) infeasible.

(c) What would result of we set $\psi = G$ in eq. (5.10) for a G making $\nabla G \cdot \mathbf{n}$ a constant on S? What must the value of this constant be?

Solution: Equation (5.15) is $\phi(0,0,0) = \frac{-1}{4\pi}\int\int\int_D \frac{\lambda^2 \phi}{R} dV + \frac{1}{4\pi}\int\int_S \left[\frac{\nabla \phi}{R} - \phi\nabla\left(\frac{1}{R}\right)\right]\cdot \mathbf{n} dS$.

(a) The second Green formula is $\int\int\int_D(\phi\nabla^2\psi - \psi\nabla^2\phi)dV = \int\int_S(\phi\nabla\psi - \psi\nabla\phi)\cdot \mathbf{n} dS$. Setting $\psi = G$ with $\nabla G \cdot \mathbf{n} = 0$ on S, we obtain $\int\int\int_D(\phi\nabla^2 G - G\nabla^2\phi)dV = -\int\int_S G\nabla\phi\cdot \mathbf{n} dS$. With $G = 1/R + \gamma(\mathbf{R})$, the laplacian of G is $\nabla^2 G = \nabla^2 \gamma(\mathbf{R}) = 0$ because γ is harmonic, so the second Green formula becomes $\int\int\int_D \left(\frac{1}{R} + \gamma\right)\nabla^2\phi dV = \int\int_S \left(\frac{1}{R}+\gamma\right)\nabla\phi\cdot\mathbf{n}dS$.

(b) For $G = \frac{1}{R} + \gamma(\mathbf{R})$, $\nabla G = -\frac{\mathbf{R}}{R^3} + \nabla\gamma(\mathbf{R})$ and $\int\int_S \nabla G\cdot \mathbf{n}dS = -\int\int_S \left(\frac{\mathbf{R}}{R^3} + \nabla\gamma(\mathbf{R})\right)\cdot \mathbf{n}dS = -\int\int_S \left(\frac{\mathbf{R}}{R^3}\right)\cdot \mathbf{n}dS + \int\int\int \nabla\cdot\nabla\gamma(\mathbf{R})dV$. Because γ is harmonic, this reduces to $-\int\int_S \left(\frac{\mathbf{R}}{R^3}\right)\cdot \mathbf{n}dS$. Letting $\phi = 1$ over the whole domain D in equation 5.15, we have $1 = 0 + \frac{1}{4\pi}\int\int_S 0 - \nabla\frac{1}{R}\cdot \mathbf{n}dS$, or $-4\pi = \int\int_S \nabla\frac{1}{R}\cdot \mathbf{n}dS = \int\int_S \left(\frac{\mathbf{R}}{R^3}\right)\cdot \mathbf{n}dS$, and the first result is proved. Because $\nabla G \cdot \mathbf{n} = 0$ on the boundary for part (a), and the evaluation of that constant over the boundary must be -4π, that function is incompatible with this result.

(c) When $\psi = G$ such that $\nabla G\cdot \mathbf{n} = C$, then the analysis of part (a) can be modified accordingly. Setting $\psi = G$ with $\nabla G\cdot \mathbf{n} = C$ on S, we obtain $\int\int\int_D(\phi\nabla^2 G - G\nabla^2\phi)dV = \int\int CdS - \int\int_S G\nabla\phi\cdot \mathbf{n}dS$. With $G = 1/R + \gamma(\mathbf{R})$, the laplacian of G is $\nabla^2 G = \nabla^2 \gamma(\mathbf{R}) = 0$ because γ is harmonic, so we have $\int\int_S \left(\frac{1}{R}+\gamma\right)\nabla\phi\cdot \mathbf{n}dS - \int\int\int_D \left(\frac{1}{R}+\gamma\right)\nabla^2\phi dV = C$ (Area), where Area is the area of the surface. Using the divergence theorem and rearranging, we have $\frac{1}{\text{Area}}\int\int\int_D \nabla\cdot\left[\left(\frac{1}{R}+\gamma\right)\nabla\phi\right] - \left(\frac{1}{R}+\gamma\right)\nabla^2\phi dV = C$ or $\frac{1}{\text{Area}}\int\int\int_D \nabla\left(\frac{1}{R}+\gamma\right)\cdot\nabla\phi + \left(\frac{1}{R}+\gamma\right)\nabla^2\phi - \left(\frac{1}{R}+\gamma\right)\nabla^2\phi dV = \frac{1}{\text{Area}}\int\int\int_D \nabla\left(\frac{1}{R}+\gamma\right)\cdot\nabla\phi dV = C$

9. With D and S as in the previous exercise, suppose that $\nabla\cdot \mathbf{V} = 0$ and $\mathbf{W} = \nabla\phi$ with $\phi = 0$ on S. Prove that $\int\int\int_D \mathbf{V}\cdot \mathbf{W}dV = 0$.
Solution: Using vector identity 3.28 $\nabla\cdot\phi\mathbf{F} = \phi\nabla\cdot\mathbf{F} + \mathbf{F}\cdot\nabla\phi$, we can rewrite $\int\int\int_D \mathbf{V}\cdot \mathbf{W}dV = \int\int\int_D \mathbf{V}\cdot\nabla\phi dV = \int\int\int_D \nabla\cdot\phi\mathbf{V} - \phi\nabla\cdot \mathbf{V}dV$. Because $\nabla\cdot \mathbf{V} = 0$, this is $\int\int\int_D \nabla\cdot\phi\mathbf{V}dV$, but ϕ is 0 on this surface, so the integral is too.

5.3 The Fundamental Theorem of Vector Analysis

1. Show that the formula (4.18) takes the form (5.30) when $\mathbf{J} = \nabla\times \mathbf{F}$.
Solution: Formula (4.18) is $\mathbf{G}(x,y,z) = \int_0^1 t\mathbf{F}\times\frac{d\mathbf{r}}{dt}dt$, and equation (5.30) is $\mathbf{F}_1(\mathbf{R}) = \int_0^t t\mathbf{J}[t(\mathbf{R}-\mathbf{R}_0)]\times[\mathbf{R}-\mathbf{R}_0]dt$. Then, following a similar derivation as the one described in the book,

$$\mathbf{F}_1(\mathbf{R}) = \int_0^t t\mathbf{J}[t(\mathbf{R}-\mathbf{R}_0)]\times[\mathbf{R}-\mathbf{R}_0]dt$$

$$\nabla\times \mathbf{F}_1(\mathbf{R}) = \int_0^t t\nabla\times(\mathbf{J}[t(\mathbf{R}-\mathbf{R}_0)]\times[\mathbf{R}-\mathbf{R}_o])\,dt =$$

$$= \int_0^1 [([\mathbf{R}-\mathbf{R}_0]\cdot\nabla)\mathbf{J}[t(\mathbf{R}-\mathbf{R}_0)] - (\mathbf{J}[t(\mathbf{R}-\mathbf{R}_0)]\cdot\nabla)[\mathbf{R}-\mathbf{R}_0] + (\nabla\cdot[\mathbf{R}-\mathbf{R}_0])\mathbf{J}[t(\mathbf{R}-\mathbf{R}_0)]$$

$$- (\nabla\cdot \mathbf{J}[t(\mathbf{R}-\mathbf{R}_0)])[\mathbf{R}-\mathbf{R}_0]]tdt$$

By the same logic used in section 4.4, $\nabla \cdot \mathbf{J} = 0$. Thus,

$$\nabla \times \mathbf{F}_1(\mathbf{R}) = \int_0^1 \left[([\mathbf{R} - \mathbf{R}_0] \cdot \nabla) \mathbf{J}[t(\mathbf{R} - \mathbf{R}_0)] - \mathbf{J}[t(\mathbf{R} - \mathbf{R}_0)] + 3\mathbf{J}[t(\mathbf{R} - \mathbf{R}_0)]\right] t\, dt \quad (5.1)$$

Then, by eq. (4.17),

$$\nabla \times \mathbf{F}_1(\mathbf{R}) = \int_0^1 \frac{d}{dt}\left(t^2 \mathbf{J}[t(\mathbf{R} - \mathbf{R}_0)]\right) dt \quad (5.2)$$

$$= t^2 \mathbf{J}[t(\mathbf{R} - \mathbf{R}_0)]\big|_0^1 = \mathbf{J}[t(\mathbf{R} - \mathbf{R}_0)] \nabla \times \mathbf{F} = \mathbf{J} \quad (5.3)$$

3. True or false: if $\nabla \cdot \mathbf{F} = 0$ and $\nabla \times \mathbf{F} = 0$, then $\mathbf{F}$ is constant.
   Solution: False, for example the divergence of $\mathbf{F} = y\mathbf{i} + x\mathbf{j}$ is zero as is the curl, but $\mathbf{F}$ is not constant.

5. Devise a proof of the fundamental theorem for star-shaped domains which begins by constructing a solenoidal field with the same curl as $\mathbf{F}$.
   Solution: The fundamental theorem of vector analysis is this: **Theorem:** Let $\mathbf{F}(\mathbf{R})$ be a continuously differentiable vector field defined in a star-shaped domain D. Then $\mathbf{F}$ can be written $\mathbf{F} = \nabla \phi + \nabla \times \mathbf{G}$ for some scalar field ϕ and some vector field $\mathbf{G}$. To prove this following our authors' suggestion, let $\mathbf{H}$ be a solenoidal field with the came curl as $\mathbf{F}$. We can construct such a solenoidal field $\mathbf{H}$ by $\mathbf{H} = \nabla \times \mathbf{G} = \frac{1}{4\pi} \int\int\int_D \frac{\nabla' \times \mathbf{F}(\mathbf{R}')}{|\mathbf{R} - \mathbf{R}'|} dV'$. Then $\mathbf{F} = \mathbf{H} + \nabla \phi$ for $\mathbf{H} = \mathbf{F} - \nabla \phi$. Because $\mathbf{H}$ is solenoidal, $\mathbf{H} = \nabla \times \mathbf{G}$ for some $\mathbf{G}$ and we have the result $\mathbf{F} = \nabla \times \mathbf{G} + \nabla \phi$.

7. Devise a counterexample to the final identity in the text. (*Hint:* There are trivial ones.)
   Solution: $\mathbf{F}$ = constant.

5.4 Green's Theorem

1. Use Green's theorem to derive equation (5.41).
   Solution: Green's theorem states $\int_C F_1 dx + F_2 dy = \int\int \left(\frac{\partial F_2}{\partial x} - \frac{\partial F_1}{\partial y}\right) dx\, dy$, the right side of which can be interpreted as an integral over a region D. If we integrate 1 over an area, we will get that area, so suppose that D has area A, and that D is convex (any line joining two points on the boundary intersects only these points on the boundary), as required by Green's theorem. Then the boundary of D will be decomposable into two regions, each of which can be parameterized by y as a function of x. The area of the region D is the integral $A = \int \int_{y=y_1(x)}^{y=y_2(x)} 1 dy\, dx = \int y_2 - y_1\, dx = \int y\, dx$. Because $1 = \frac{\partial F_2}{\partial x} - \frac{\partial F_1}{\partial y}$ when $F_2 = 0$ and $F_1 = y$, we have $\int\int \left(\frac{\partial F_2}{\partial x} - \frac{\partial F_1}{\partial y}\right) dy\, dx = \int\int \left(\frac{\partial 0}{\partial x} - \frac{\partial y}{\partial y}\right) dy\, dx = -\int 1 dx = -A$.

3. Use Green's theorem to derive equation (5.43).
   Solution: Let $F_1 = x$ and $F_2 = 0$. Then $\int\int \left(\frac{\partial F_2}{\partial x} - \frac{\partial F_1}{\partial y}\right) dx\, dy = \int\int \left(\frac{\partial 0}{\partial x} - \frac{\partial x}{\partial y}\right) dx\, dy = 0 = \int x\, dx$ by Green's theorem.

5. Let $\mathbf{R} = x\mathbf{i} + y\mathbf{j}$ and $d\mathbf{R} = dx\mathbf{i} + dy\mathbf{j}$.

   (a) Compute the magnitude of the vector cross product $\mathbf{R} \times (\mathbf{R} + d\mathbf{R})$.

   (b) Thus, give a direct geometrical interpretation of the integrand of eq. (5.49). [*Hint:* Consider the triangle with the following vertices: $(0,0)$, (x,y), and $(x+dx, y+dy)$.]

   (c) Using figure 5.6, give an alternative derivation of eq. (5.49).

   Solution:

5.4. GREEN'S THEOREM

(a) $\mathbf{R} \times (\mathbf{R} + d\mathbf{R}) = \begin{vmatrix} \mathbf{i} & \mathbf{j} & \mathbf{k} \\ x & y & 0 \\ x+dx & y+dy & 0 \end{vmatrix} = (xdy - ydx)\mathbf{k}$. The magnitude is $\sqrt{(xdy-ydx)^2} = |xdy - ydx|$

(b) Equation (5.49) states $\int_C \frac{1}{2}(x\,dy - y\,dx) = A$. Now, $\int_C \frac{1}{2}(x\,dy - y\,dx) = \int |\mathbf{R} \times (\mathbf{R} + d\mathbf{R})|$, and this is the area of a parallelogram, so one half of this is the area of the triangle given. Refer now to part c.

(c) Subtracting the area of the triangle $(0,0)$, (x,y), $(x+dx, y+dy)$ on the lower curve from that of the triangle $(0,0)$, (x,y), $(\hat{x}+d\hat{x}, \hat{y}+d\hat{y})$ on the upper curve, we get the area of the wedge in the interior of the region in Figure 5.6. Adding these all up and taking the limit as dx and dx go to zero gives us the area of the figure.

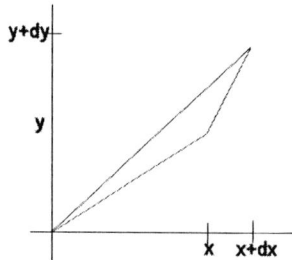

Figure 5.2: The triangle $(0,0)$, (x,y), $(x+dx, y+dy)$.

7. Let C denote the circle $x^2 + y^2 = 9$, and let $\mathbf{F} = y\mathbf{i} - 3x\mathbf{j}$. What is the line integral of the tangential component of $\mathbf{F}$ around C, taken in the usual counterclockwise direction?
   Solution: We can compute the integral around the circle by integrating the flux of the curl of the vector field through any simple surface bounded by this circle. Choosing the simplest to be the disk D with outward normal $\mathbf{k}$, we find $\int \int_D \nabla \times \mathbf{F} n dS = \int \int_D \nabla \times (y\mathbf{i} - 3x\mathbf{j}) \cdot \mathbf{k} dS = \int \int_D -4 dS$. Because the area of the disk is $\pi r^2 = \pi 3^2 = 9\pi$, the integral is -36π.

9. Compute $\int_C (4y^3 dx - 2x^2 dy)$ around the square bounded by the lines $x = \pm 1$ and $y = \pm 1$,

   (a) directly, by performing the line integration.
   (b) by using Green's theorem.

   By symmetry, it is obvious that one of the terms in the integrand of the above line integral can be ignored. Which term?
   Solution:

   (a) To calculate the line integral directly, we begin by finding the parametrization of the four lines. The line segment from $(1, -1, 0)$ to $(1, 1, 0)$ can be written as $S_1 = \mathbf{i} + y\mathbf{j}$. The other three lines can be written as $S_2 = x\mathbf{i} + \mathbf{j}$, $S_3 = -\mathbf{i} + y\mathbf{j}$ and $S_4 = x\mathbf{i} - \mathbf{j}$. We can then find dR for each line. $dR_1 = \mathbf{j}$, $dR_2 = \mathbf{i}$, $dR_3 = \mathbf{j}$ and $dR_4 = \mathbf{i}$. Thus, the line integral for the first line is $\int_{-1}^{1} F \cdot dR_1 = \int_{-1}^{1}(4y^3\mathbf{i} - 2\mathbf{j}) \cdot \mathbf{j} dy = \int_{-1}^{1} -2dy = -4$. The line integral for the second line is $\int_1^{-1}(4\mathbf{i} - 2x^2\mathbf{j}) \cdot \mathbf{i} = -8$. The line integral for the third line is $\int_1^{-1}(4y^3\mathbf{i} + 2\mathbf{j}) \cdot \mathbf{j} = 4$. And, finally, the line integral for the fourth line segment is $\int_{-1}^{1}(-4\mathbf{i} - 2x^2\mathbf{j}) \cdot \mathbf{i} = -8$. So, the sum of these line integrals is the line integral over the boundary of the square and is equal to $-8 + 4 - 4 - 8 = -16$.

   (b) Using Green's theorem, $\int_C (4y^3 dx - 2x^2 dy) = \int \int \left(\frac{\partial F_2}{\partial x} - \frac{\partial F_1}{\partial y} \right) dS$. $\int \int \left(\frac{\partial F_2}{\partial x} - \frac{\partial F_1}{\partial y} \right) dS = \int \int (-4x - 12y^2) dS = \int_{-1}^{1} \int_{-1}^{1} (-4x - 12y^2) dxdy = \int_{-1}^{1} (-24y^2) dy = -16$

   The second term can be ignored because $-2x^2$ is an even function and its integral along the top and bottom of the square cancel.

11. In eq. (5.50), the functions F_1 and F_2 are fairly arbitrary functions of x and y (we require only that certain partial derivatives be continuous). It therefore appears that we can interchange F_1 and F_2

and also x and y to obtain the formula $\int_C (F_2 dy + F_1 dx) = \int\int_D \left(\frac{\partial F_1}{\partial dy} - \frac{\partial F_2}{\partial x}\right) dy dx$. The left side of this equation is the same as the left side of equation (5.50), but the right side has the opposite sign. It follows that this expression is incorrect. Give a clue, *in only one word*, to explain this paradox.

Solution: Orientation. Green's theorem in the plane can be interpreted as the relation between the flux of the curl of a vector field $\mathbf{F} = F_1\mathbf{i} + F_2\mathbf{j} + F_3\mathbf{k}$ through a closed planar region $\int\int \nabla\times\mathbf{F}\cdot d\mathbf{S}$ and the line integral of $\mathbf{F}$ over the boundary $\int \mathbf{F}\cdot d\mathbf{R}$. In the xy plane, $d\mathbf{S} = \mathbf{k} dy dx$, so Green's theorem is $\int\int \left(\frac{\partial F_2}{\partial x} - \frac{\partial F_1}{\partial y}\right) dy dx = \int \mathbf{F}\cdot d\mathbf{R}|_{z=0} = \int F_1 dx + F_2 dy$. Reversing the order $\int F_2 dy + F_1 dx$ violates the spirit of interpretation of the sum as the ordered components of a dot product with zero z component, and reverses the order of the vector components in the $\mathbf{i}$ and $\mathbf{j}$ directions, reversing the orientation around the boundary curve.

5.5 Stokes' Theorem

1. Given the vector field $\mathbf{F} = 3y\mathbf{i} + (5-2x)\mathbf{j} + (z^2-2)\mathbf{k}$, find

   (a) div $\mathbf{F}$.

   (b) **curlF**.

   (c) the surface integral of the normal component of **curlF** over the open hemispherical surface $x^2 + y^2 + z^2 = 4$ above the xy plane. [*Hint:* By a double application of Stokes' theorem, part (c) can be reduced to a triviality.]

   Solution:

   (a) The divergence is $\nabla\cdot\mathbf{F} = \nabla\cdot(3y\mathbf{i} + (5-2x)\mathbf{j} + (z^2-2)\mathbf{k}) = 0 + 0 + 2z$.

   (b) The curl is $\nabla\times(3y\mathbf{i} + (5-2x)\mathbf{j} + (z^2-2)\mathbf{k}) = \begin{vmatrix} \mathbf{i} & \mathbf{j} & \mathbf{k} \\ \partial_x & \partial_y & \partial_z \\ 3y & 5-2x & z^2-2 \end{vmatrix} = (-2-3)\mathbf{k} = -5\mathbf{k}$.

   (c) By an application of Stoke's theorem, we see that rather than computing the curl of $\mathbf{F}$ over the hemisphere, we can compute the line integral of $\mathbf{F}$ over the boundary circle $x^2 + y^2 = 4$. However, the integral over the bounding circle of the tangential component of $\mathbf{F}$ is the same (by the second application of Stokes' theorem) as integrating the curl of $\mathbf{F}$ over the disc in the xy plane bounded by the circle. Therefore, we calculate $\int\int \nabla\times\mathbf{F}\cdot d\mathbf{S} = \int\int -5\mathbf{k}\cdot\mathbf{k} dS = -5\int\int dS = -5(4\pi) = -20\pi$.

3. Prove $\int\int_S \nabla\phi\times\nabla\psi\cdot d\mathbf{S} = \oint_C \phi\nabla\psi\cdot d\mathbf{R}$.

   Solution: If we can show that $\nabla\times(\phi\nabla\psi) = \nabla\phi\times\nabla\psi$, then we will be done, because the equation $\int\int_S \nabla\phi\times\nabla\psi\cdot d\mathbf{S} = \oint_C \phi\nabla\psi\cdot d\mathbf{R}$ becomes a case of Stokes' theorem. From vector identity 3.29 on page 147, $\nabla\times(\phi\mathbf{F}) = \phi\nabla\times\mathbf{F} + \nabla\phi\times\mathbf{F}$, and with $\mathbf{F} = \nabla\psi$, this becomes $\nabla\times(\phi\nabla\psi) = \phi\nabla\times\nabla\psi + \nabla\phi\times\nabla\psi$. But by vector identity 3.40, $\nabla\times\nabla(\phi) = 0$, so $\nabla\times(\phi\nabla\psi) = \nabla\phi\times\nabla\psi$, which we wished to prove.

5. Be a bit fanciful, and imagine that S is the surface of a laundry bag with a drawstring forming the boundary C. The Stokes' theorem states that the normal component of the surface integral of **curlF** over the laundry bag equals the line integral of its tangential component around the drawstring. Now suppose that we close the laundry bag by pulling the drawstring; the effective length of the drawstring becomes zero and the line integral is therefore zero. S has become a closed surface.

   (a) What is the surface integral of the normal component of **curlF** over a closed surface?

   We now apply the divergence theorem, which says that the volume integral of of the divergence of a vector field through the interior of a closed laundry bag equals the surface integral of the normal component of the field over its surface. Let the vector field be **curlF**.

5.6. THE TRANSPORT THEOREMS

(b) What is the volume integral of the divergence of **curlF** over a domain?

If the laundry bag is very, very small, the divergence of **curlF** will be approximately constant throughout, and the volume integral of div **curlF** will be approximately div **curlF** at a point within the laundry bag times the volume that the bag encloses.

(c) What is div **curlF** at any point P?

(d) To which of the identities in 3.8 is this related?

Solution:

(a) Because the integral of **F** around the hole goes to zero in the limit as the length of the boundary of the hole goes to zero, the flux of the curl of **F** goes to zero also.

(b) The divergence of the curl is always 0.

(c) 0.

(d) div **curlF** $= 0$.

7. Suppose that $\mathbf{F} = \mathbf{grad}\phi$, so that the line integral of the tangential component of **F** along any curve is equal to the difference in the values of ϕ at the endpoints of the curve. In particular, if C is a closed curve, $\int_C \mathbf{F} \cdot d\mathbf{R} = 0$. Let S be a surface with boundary C

   (a) What is the surface integral of the normal component of **curlgrad**ϕ over a surface S?

   If S is a small element of surface, bounded by a closed curve C, **curlgrad**ϕ will be approximately constant on S, and the surface integral of the normal component of **curlgrad**ϕ will be approximately $\mathbf{n} \cdot$ **curlgrad**ϕ times the area of the surface.

   (b) For any unit vector **n**, and any point in space, what is $\mathbf{n} \cdot$ **curlgrad**ϕ at this point?

   (c) Since this result is independent of the direction of **n**, what can you say about **curlgrad**ϕ?

   (d) To which of the identities in section 3.8 is this related?

Solution:

(a) The curl of a gradient is zero: $\begin{vmatrix} \mathbf{i} & \mathbf{j} & \mathbf{k} \\ \partial_x & \partial_y & \partial_z \\ \frac{\partial \phi}{\partial x} & \frac{\partial \phi}{\partial y} & \frac{\partial \phi}{\partial z} \end{vmatrix} = 0$, so the integral over any surface is zero.

(b) For the same reason as in part (a) the result is 0.

(c) Again, for the same reason as in part (a), the curl of the gradient is the zero vector **0**.

(d) These are all related to the fact that **curlgrad**$\phi = \mathbf{0}$.

5.6 The Transport Theorems

1. With a rigorous proof of Reynolds' theorem in hand, revisit exercise 12 of section 3.3 by proving Euler's expansion formula:

$$\frac{d}{dt} \iiint_{D(t)} dV = \frac{d\,(\text{volume})}{dt} \iiint_{D(t)} \nabla \cdot \mathbf{v}\, dV = \iint_S \mathbf{v} \cdot d\mathbf{S}.$$

Solution: Exercise 12 of 3.3 states: "Another hydrodynamic interpretation of divergence is as follows. Let **F** be the velocity field of a fluid. Consider a small rectangular parallelepiped of fluid located at (x, y, z). Then the divergence of **F** is the time rate of change of volume of this body of fluid, per unit volume, as the size of the box goes to zero. Show this."

Reynold's theorem states, for a time-varying scalar field $\eta(\mathbf{R}, t)$ and time varying domain $D(t)$,

$$\frac{d}{dt} \iiint_{D(t)} dV = \iiint_{D(t)} \left\{ \frac{d\eta}{dt} + \nabla \cdot \eta\mathbf{v} \right\} dV = \iiint_{D(t)} \frac{d\eta}{dt} dV + \iint_{S(t)} \eta\mathbf{v} \cdot d\mathbf{S}.$$

Letting $\eta(\mathbf{R}, t) = 1$, we obtain the result immediately.

5.7 Matrix Techniques in Vector Analysis

1. Form the indicated products

(a) $\begin{bmatrix} 1 & 2 & 3 \end{bmatrix} \begin{bmatrix} 1 \\ 2 \\ 3 \end{bmatrix}$

(b) $\begin{bmatrix} 1 & 2 & 3 \end{bmatrix} \begin{bmatrix} 1 & 0 & -1 \\ 0 & 1 & 1 \\ 0 & -1 & 2 \end{bmatrix}$

(c) $\begin{bmatrix} 1 & 0 & -1 \\ 0 & 1 & 1 \\ 0 & -1 & 2 \end{bmatrix} \begin{bmatrix} 1 \\ 2 \\ 3 \end{bmatrix}$

(d) $\begin{bmatrix} 1 & 2 & 3 \end{bmatrix} \begin{bmatrix} 1 & 0 & -1 & 3 \\ 0 & 1 & 1 & 2 \\ 0 & -1 & 2 & 1 \end{bmatrix}$

(e) $\begin{bmatrix} 1 & 2 & 3 \\ 2 & 1 & 1 \\ 1 & 2 & 3 \end{bmatrix} \begin{bmatrix} 1 & 0 & -1 \\ 0 & 1 & 1 \\ 0 & -1 & 2 \end{bmatrix}$

(f) $\begin{bmatrix} 1 & 0 & -1 \\ 0 & 1 & 1 \\ 0 & -1 & 2 \end{bmatrix} \begin{bmatrix} 10 & 0 & 0 \\ 0 & 10 & 0 \\ 0 & 0 & 10 \end{bmatrix}$

Solution:

(a) $\begin{bmatrix} 1 & 2 & 3 \end{bmatrix} \begin{bmatrix} 1 \\ 2 \\ 3 \end{bmatrix} = 1 \cdot 1 + 2 \cdot 2 + 3 \cdot 3 = 14.$

(b) $\begin{bmatrix} 1 & 2 & 3 \end{bmatrix} \begin{bmatrix} 1 & 0 & -1 \\ 0 & 1 & 1 \\ 0 & -1 & 2 \end{bmatrix} =$
$= \begin{bmatrix} 1 \cdot 1 + 2 \cdot 0 + 3 \cdot 0 & 1 \cdot 0 + 2 \cdot 1 + 3 \cdot -1 & 1 \cdot -1 + 2 \cdot 1 + 3 \cdot 2 \end{bmatrix} =$
$= \begin{bmatrix} 1 & -1 & 7 \end{bmatrix}.$

(c) $\begin{bmatrix} 1 & 0 & -1 \\ 0 & 1 & 1 \\ 0 & -1 & 2 \end{bmatrix} \begin{bmatrix} 1 \\ 2 \\ 3 \end{bmatrix} = \begin{bmatrix} 1 \cdot 1 + 0 \cdot 2 - 1 \cdot 3 \\ 0 \cdot 1 + 1 \cdot 2 + 1 \cdot 3 \\ 0 \cdot 1 - 1 \cdot 2 + 2 \cdot 3 \end{bmatrix} = \begin{bmatrix} -2 \\ 5 \\ 4 \end{bmatrix}.$

(d) $\begin{bmatrix} 1 & 2 & 3 \end{bmatrix} \begin{bmatrix} 1 & 0 & -1 & 3 \\ 0 & 1 & 1 & 2 \\ 0 & -1 & 2 & 1 \end{bmatrix} =$
$= \begin{bmatrix} 1 \cdot 1 + 2 \cdot 0 + 3 \cdot 0 & 1 \cdot 0 + 2 \cdot 1 + 3 \cdot -1 & 1 \cdot -1 + 2 \cdot 1 + 3 \cdot 2 & 1 \cdot 3 + 2 \cdot 2 + 3 \cdot 1 \end{bmatrix} =$
$= \begin{bmatrix} 1 & -1 & 7 & 10 \end{bmatrix}.$

(e) $\begin{bmatrix} 1 & 2 & 3 \\ 2 & 1 & 1 \\ 1 & 2 & 3 \end{bmatrix} \begin{bmatrix} 1 & 0 & -1 \\ 0 & 1 & 1 \\ 0 & -1 & 2 \end{bmatrix} =$
$= \begin{bmatrix} 1 \cdot 1 + 2 \cdot 0 + 3 \cdot 0 & 1 \cdot 0 + 2 \cdot 1 + 3 \cdot -1 & 1 \cdot -1 + 2 \cdot 1 + 3 \cdot 2 \\ 2 \cdot 1 + 1 \cdot 0 + 1 \cdot 0 & 2 \cdot 0 + 1 \cdot 1 + 1 \cdot -1 & 2 \cdot -1 + 1 \cdot 1 + 1 \cdot 2 \\ 1 \cdot 1 + 2 \cdot 0 + 3 \cdot 0 & 1 \cdot 0 + 2 \cdot 1 + 3 \cdot -1 & 1 \cdot -1 + 2 \cdot 1 + 3 \cdot 2 \end{bmatrix} = \begin{bmatrix} 1 & -1 & 7 \\ 2 & 0 & 1 \\ 1 & -1 & 7 \end{bmatrix}.$

(f) $\begin{bmatrix} 1 & 0 & -1 \\ 0 & 1 & 1 \\ 0 & -1 & 2 \end{bmatrix} \begin{bmatrix} 10 & 0 & 0 \\ 0 & 10 & 0 \\ 0 & 0 & 10 \end{bmatrix} =$
$= \begin{bmatrix} 1 \cdot 10 + 0 \cdot 0 - 1 \cdot 0 & 1 \cdot 0 + 0 \cdot 10 - 1 \cdot 0 & 1 \cdot 0 + 0 \cdot 0 - 1 \cdot 10 \\ 0 \cdot 10 + 1 \cdot 0 + 1 \cdot 0 & 0 \cdot 0 + 1 \cdot 10 + 1 \cdot 0 & 0 \cdot 0 + 1 \cdot 0 + 1 \cdot 10 \\ 0 \cdot 10 - 1 \cdot 0 + 2 \cdot 0 & 0 \cdot 0 - 1 \cdot 10 + 2 \cdot 0 & 0 \cdot 0 - 1 \cdot 0 + 2 \cdot 10 \end{bmatrix} = \begin{bmatrix} 10 & 0 & -10 \\ 0 & 10 & 10 \\ 0 & -10 & 20 \end{bmatrix}.$

3. Prove the distributive laws in eq. (5.87).
Solution:
Show that $(s\mathbf{M} + \mathbf{N})\mathbf{R} = s\mathbf{MR} + \mathbf{NR}$.

$((s\mathbf{M} + \mathbf{N})\mathbf{R})_{ij} = \sum_{k=1}^{n}(s\mathbf{M}_{ik} + \mathbf{N}_{ik})\mathbf{R}_{kj}$

$= \sum_{k=1}^{n}(s\mathbf{M}_{ik}\mathbf{R}_{kj} + \mathbf{N}_{ik}\mathbf{R}_{kj})$

$= \sum_{k=1}^{n}(s\mathbf{M}_{ik}\mathbf{R}_{kj}) + \sum_{k=1}^{n}(\mathbf{N}_{ik}\mathbf{R}_{kj})$

$= s\mathbf{MR} + \mathbf{NR}$

Show that $\mathbf{R}(s\mathbf{M} + \mathbf{N}) = s\mathbf{RM} + \mathbf{RN}$.

5.7. MATRIX TECHNIQUES IN VECTOR ANALYSIS

$$\begin{aligned}(\mathbf{R}(s\mathbf{M}+\mathbf{N}))_{ij} &= \sum_{k=1}^{n} \mathbf{R}_{ik}(s\mathbf{M}_{kj} + \mathbf{N}_{kj}) \\ &= \sum_{k=1}^{n}(s\mathbf{R}_{ik}\mathbf{M}_{kj} + \mathbf{R}_{ij}\mathbf{N}_{kj}) \\ &= \sum_{k=1}^{n}(s\mathbf{R}_{ik}\mathbf{M}_{kj}) + \sum_{k=1}^{n}(\mathbf{R}_{ik}\mathbf{N}_{kj}) \\ &= s\mathbf{RM} + \mathbf{RN}\end{aligned}$$

5. Show that if **A**, **B** and **C** are coplanar, then no reciprocal set of vectors exists. (Do not forget to consider the possibility that they are colinear.)
Solution: A set of vectors **D**, **E** and **F** reciprocal to **A**, **B** and **C** would have the property $\mathbf{A} \cdot \mathbf{D} = \mathbf{B} \cdot \mathbf{E} = \mathbf{C} \cdot \mathbf{F} = 1$ and $\mathbf{A} \cdot \mathbf{E} = \mathbf{A} \cdot \mathbf{F} = 0$, $\mathbf{B} \cdot \mathbf{D} = \mathbf{B} \cdot \mathbf{F} = 0$, $\mathbf{C} \cdot \mathbf{E} = \mathbf{C} \cdot \mathbf{D} = 0$. This can be easily encoded in the matrix equation $AD = I$, specifically,

$$\begin{bmatrix} \cdots & \mathbf{A} & \cdots \\ \cdots & \mathbf{B} & \cdots \\ \cdots & \mathbf{C} & \cdots \end{bmatrix} \begin{bmatrix} \vdots & \vdots & \vdots \\ \mathbf{D} & \mathbf{E} & \mathbf{F} \\ \vdots & \vdots & \vdots \end{bmatrix} = \begin{bmatrix} 1 & 0 & 0 \\ 0 & 1 & 0 \\ 0 & 0 & 1 \end{bmatrix}.$$

A solution $A^{-1} = D$ exists only if a left inverse for A exists, which by equation (5.58) requires that $[\mathbf{A}, \mathbf{B}, \mathbf{C}]$ be nonzero. Because **A**, **B** and **C** are coplanar, $[\mathbf{A}, \mathbf{B}, \mathbf{C}] = 0$ and no reciprocal set of vectors exists.

7. Show that the right inverse in eq. (5.92) agrees with the formula (5.97).
Solution: Equation 5.92 is $\begin{bmatrix} 1 & -2 & 1 \\ 2 & 0 & 1 \\ 1 & 1 & 2 \end{bmatrix} \begin{bmatrix} -\frac{1}{4} & \frac{3}{4} & -\frac{1}{4} \\ -\frac{3}{4} & \frac{1}{4} & \frac{1}{4} \\ \frac{1}{2} & -\frac{1}{2} & -\frac{1}{2} \end{bmatrix} = \begin{bmatrix} 1 & 0 & 0 \\ 0 & 1 & 0 \\ 0 & 0 & 1 \end{bmatrix}$. Formula (5.97) says that the inverse of $M = \begin{bmatrix} \cdots & \mathbf{A} & \cdots \\ \cdots & \mathbf{B} & \cdots \\ \cdots & \mathbf{C} & \cdots \end{bmatrix}$ is given by $R = \frac{1}{[\mathbf{A},\mathbf{B},\mathbf{C}]} \begin{bmatrix} \vdots & \vdots & \vdots \\ \mathbf{B} \times \mathbf{C} & \mathbf{C} \times \mathbf{A} & \mathbf{A} \times \mathbf{B} \\ \vdots & \vdots & \vdots \end{bmatrix}$.

We find $\frac{1}{[\mathbf{A},\mathbf{B},\mathbf{C}]} = \frac{1}{4}$, $\mathbf{A} \times \mathbf{B} = -\mathbf{i} + \mathbf{j} + 2\mathbf{k}$, $\mathbf{B} \times \mathbf{C} = -\mathbf{i} - 3\mathbf{j} + 2\mathbf{k}$, and $\mathbf{C} \times \mathbf{A} = 3\mathbf{i} + \mathbf{j} - 2\mathbf{k}$, so $R = \frac{1}{4}\begin{bmatrix} -1 & 3 & -1 \\ -3 & 1 & 1 \\ 2 & -2 & 2 \end{bmatrix}$, which is the same as the right inverse in equation 5.92.

9. Prove that if $\det M \neq 0$, M has only one inverse.
Solution: Suppose $|M| \neq 0$. Then by equations (5.97) and (5.98) left and right inverses exist, that is, there exist matrices P and Q such that $PM = I$ and $MQ = I$. Multiply, say, $PM = I$ on the right by Q: $PMQ = IQ \Rightarrow PI = IQ \Rightarrow P = Q$, so if the determinant of M is not zero, there is a unique inverse.

11. Solve exercise 10 by Cramer's rule.
Solution: Exercise 10 written in matrix form is $M\mathbf{x} = \mathbf{a}$ or $\begin{bmatrix} 2 & 1 & 2 \\ 3 & 0 & 2 \\ 1 & 1 & 2 \end{bmatrix} \begin{bmatrix} x \\ y \\ z \end{bmatrix} = \begin{bmatrix} 2 \\ 4 \\ 0 \end{bmatrix}$.

Cramer's rule lets us compute the components of $\begin{bmatrix} x \\ y \\ z \end{bmatrix}$ by computing ratios of determinants derived from M and $\mathbf{a}$. We compute $x = \frac{\begin{vmatrix} 2 & 1 & 2 \\ 4 & 0 & 2 \\ 0 & 1 & 2 \end{vmatrix}}{\begin{vmatrix} 2 & 1 & 2 \\ 3 & 0 & 2 \\ 1 & 1 & 2 \end{vmatrix}} = 2$, $y = \frac{\begin{vmatrix} 2 & 2 & 2 \\ 3 & 4 & 2 \\ 1 & 0 & 2 \end{vmatrix}}{\begin{vmatrix} 2 & 1 & 2 \\ 3 & 0 & 2 \\ 1 & 1 & 2 \end{vmatrix}} = 0$, $z = \frac{\begin{vmatrix} 2 & 1 & 2 \\ 3 & 0 & 4 \\ 1 & 1 & 0 \end{vmatrix}}{\begin{vmatrix} 2 & 1 & 2 \\ 3 & 0 & 2 \\ 1 & 1 & 2 \end{vmatrix}} = -1$, where the column vector $\begin{bmatrix} 2 \\ 4 \\ 0 \end{bmatrix}$ replaces the first, second and third columns of M in the numerators of the computation of x, y and z successively. Thus $\mathbf{x} = 2\mathbf{i} - \mathbf{k}$.

13. Show that the (j,i)th entry in M^T equals the (i,j)th entry in M.
Solution: By exchanging rows for columns, as the transpose does, we exchange the row index for the column index.

15. Show that a symmetric matrix must have the form
$$\begin{bmatrix} m_{11} & m_{21} & m_{31} \\ m_{21} & m_{22} & m_{32} \\ m_{31} & m_{32} & m_{33} \end{bmatrix}$$

Is the product of two symmetric matrices symmetric?
Solution: By definition a symmetric matrix is the same as its transpose, so the matrix given is symmetric by definition. However, not every matrix is 3 by 3, so this is *not* the general form of a symmetric matrix, as any n by n matrix that equals its transpose is symmetric. Now consider the symmetric matrices M and N. If their product is symmetric, then $MN = M^T N^T = (MN)^T = N^T M^T$, so if M and N commute, that is, $MN = NM$, then their product is symmetric. On the other hand, consider, for example, the matrix products $AB = \begin{bmatrix} a & b \\ b & c \end{bmatrix}\begin{bmatrix} d & e \\ e & f \end{bmatrix} = \begin{bmatrix} ad+be & ae+bf \\ bd+ce & be+cf \end{bmatrix}$ and $BA = \begin{bmatrix} d & e \\ e & f \end{bmatrix}\begin{bmatrix} a & b \\ b & c \end{bmatrix} = \begin{bmatrix} ad+be & bd+ce \\ ae+bf & be+cf \end{bmatrix}$. Even though A and B are symmetric, it is apparent that they do not commute, so although commuting matrices must be symmetric, it is not so that symmetric matrices must commute.

17. Show that if O is orthogonal and S is symmetric, then $O^{-1}SO$ is symmetric.
Solution: A matrix O is orthogonal if $OO^T = I$ and a matrix S is symmetric if $S = S^T$. So, by rearranging the definition of an orthogonal matrix, we find that $O^{-1} = O^T$. Substituting this into the original equation gives $O^{-1}SO = O^T SO$. If we take the transpose of this equation, we see that $(O^T SO)^T = O^T S^T O$ (since $(AB)^T = B^T A^T$) and, from the symmetric property of S, $(O^T SO)^T = O^T SO$. So, by the definition of a symmetric matrix, $O^{-1}SO$ is symmetric.

19. Construct examples of systems of equations of the form of eq. (5.99) whose solutions constitute

(a) a plane.

(b) a straight line.

(c) the empty set.

(*Hint:* Remember the interpretation as the intersection of three planes.) What is the determinant of the matrix in these cases?
Solution:

(a) $\begin{bmatrix} 1 & 0 & 0 \\ 0 & 0 & 0 \\ 0 & 0 & 0 \end{bmatrix}\begin{bmatrix} x \\ y \\ z \end{bmatrix} = \begin{bmatrix} 1 \\ 0 \\ 0 \end{bmatrix}$, so $x = 1$ and y and z are arbitrary, which describes a plane through $x = 1$ and parallel to the yz plane.

(b) $\begin{bmatrix} 1 & 0 & 0 \\ 0 & 0 & 0 \\ 0 & 0 & 1 \end{bmatrix}\begin{bmatrix} x \\ y \\ z \end{bmatrix} = \begin{bmatrix} 1 \\ 0 \\ 1 \end{bmatrix}$, so $x = 1$, $z = 1$ and y is arbitrary, which describes a line.

(c) $\begin{bmatrix} 1 & 0 & 1 \\ 0 & 0 & 0 \\ 1 & 0 & -1 \end{bmatrix}\begin{bmatrix} x \\ y \\ z \end{bmatrix} = \begin{bmatrix} 1 \\ 1 \\ 1 \end{bmatrix}$, so we have both $x + z = 1$ and $x - z = 1$ with y arbitrary, which has no solutions.

The determinants in all these cases are 0.

5.8 Linear Orthogonal Transformations

1. Show that eq. (5.107) can be derived by taking scalar products of eq. (5.101) with $\mathbf{i}'$, $\mathbf{j}'$ and $\mathbf{k}'$ in turn.
Solution: Equation (5.101) states $\mathbf{R} = x\mathbf{i} + y\mathbf{j} + z\mathbf{k} = x'\mathbf{i}' + y'\mathbf{j}' + z'\mathbf{k}'$, and equation (5.107) states that $\begin{bmatrix} x' \\ y' \\ z' \end{bmatrix} = \begin{bmatrix} \mathbf{i}'\cdot\mathbf{i} & \mathbf{i}'\cdot\mathbf{j} & \mathbf{i}'\cdot\mathbf{k} \\ \mathbf{j}'\cdot\mathbf{i} & \mathbf{j}'\cdot\mathbf{j} & \mathbf{j}'\cdot\mathbf{k} \\ \mathbf{k}'\cdot\mathbf{i} & \mathbf{k}'\cdot\mathbf{j} & \mathbf{k}'\cdot\mathbf{k} \end{bmatrix}\begin{bmatrix} x \\ y \\ z \end{bmatrix}$. Taking scalar products of $\mathbf{R} = x\mathbf{i} + y\mathbf{j} + z\mathbf{k} = $

5.8. LINEAR ORTHOGONAL TRANSFORMATIONS

$x'\mathbf{i'} + y'\mathbf{j'} + z'\mathbf{k'}$ with $\mathbf{i'}$, $\mathbf{j'}$ and $\mathbf{k'}$ we find $x\mathbf{i}\mathbf{i'} + y\mathbf{j}\mathbf{i'} + z\mathbf{k}\mathbf{i'} = x'$, $x\mathbf{i}\mathbf{j'} + y\mathbf{j}\mathbf{j'} + z\mathbf{k}\mathbf{j'} = y'$, and $x\mathbf{i}\mathbf{k'} + y\mathbf{j}\mathbf{k'} + z\mathbf{k}\mathbf{k'} = z'$, which are the dot products of the rows of the matrix with the column

matrix $\begin{bmatrix} x \\ y \\ z \end{bmatrix}$.

3. Verify that the transpose of J in eq. (5.104) equals its inverse.

Solution: $J = \begin{bmatrix} \cos\theta & -\sin\theta & 0 \\ \sin\theta & \cos\theta & 0 \\ 0 & 0 & 1 \end{bmatrix}$ and $J^T = \begin{bmatrix} \cos\theta & \sin\theta & 0 \\ -\sin\theta & \cos\theta & 0 \\ 0 & 0 & 1 \end{bmatrix}$

If $J^T = J^{-1}$ then $JJ^T = I$.

$$JJ^T = \begin{bmatrix} \cos\theta & -\sin\theta & 0 \\ \sin\theta & \cos\theta & 0 \\ 0 & 0 & 1 \end{bmatrix} \begin{bmatrix} \cos\theta & \sin\theta & 0 \\ -\sin\theta & \cos\theta & 0 \\ 0 & 0 & 1 \end{bmatrix}$$

$$= \begin{bmatrix} \cos^2\theta + \sin^2\theta + 0 & \sin\theta\cos\theta - \sin\theta\cos\theta + 0 & 0+0+0 \\ \sin\theta\cos\theta - \sin\theta\cos\theta + 0 & \cos^2\theta + \sin^2\theta + 0 & 0+0+0 \\ 0+0+0 & 0+0+0 & 0+0+1 \end{bmatrix}$$

$$= \begin{bmatrix} 1 & 0 & 0 \\ 0 & 1 & 0 \\ 0 & 0 & 1 \end{bmatrix}$$

$= I$

Therefore, $J^T = J^{-1}$.

5. Consider the scalar and vector fields $f(x,y,z) = xyz$, $\mathbf{V}(x,y,z) = xz\mathbf{i} + \mathbf{j} + xyz\mathbf{k}$.
If the coordinate transformation of example 5.6 is performed with $\theta = \pi/6$, express the following fields in the new coordinate system:

(a) the scalar field f.

(b) the vector field $\mathbf{V}$.

(c) $\mathbf{grad} f$.

(d) div $\mathbf{V}$.

(e) $\mathbf{curl V}$.

Solution:

(a) Write x, y and z in terms of x', y' and z':

$$\begin{bmatrix} x \\ y \\ z \end{bmatrix} = \begin{bmatrix} \cos\pi/6 & -\sin\pi/6 & 0 \\ \sin\pi/6 & \cos\pi/6 & 0 \\ 0 & 0 & 1 \end{bmatrix} \begin{bmatrix} x' \\ y' \\ z' \end{bmatrix} = \begin{bmatrix} \frac{\sqrt{3}}{2} & -\frac{1}{2} & 0 \\ \frac{1}{2} & \frac{\sqrt{3}}{2} & 0 \\ 0 & 0 & 1 \end{bmatrix} \begin{bmatrix} x' \\ y' \\ z' \end{bmatrix} = \begin{bmatrix} \frac{\sqrt{3}}{2}x' - \frac{1}{2}y' \\ \frac{1}{2}x' + \frac{\sqrt{3}}{2}y' \\ z' \end{bmatrix}.$$

Then in the new coordinates $f(x,y,z) = xyz$ transforms to $f'(x',y',z') = (\frac{\sqrt{3}}{2}x' - \frac{1}{2}y')(\frac{1}{2}x' + \frac{\sqrt{3}}{2}y')z'$

(b) Using the results of exercise (5a) to find the components of $\mathbf{V}$ in terms of x', y' and z', we compute the field in the new coordinate basis $\mathbf{i'}, \mathbf{j'}, \mathbf{k'}$ by

$$\mathbf{V'}(x',y',z') = \begin{bmatrix} \frac{\sqrt{3}}{2} & \frac{1}{2} & 0 \\ -\frac{1}{2} & \frac{\sqrt{3}}{2} & 0 \\ 0 & 0 & 1 \end{bmatrix} \begin{bmatrix} z'(\frac{\sqrt{3}}{2}x' - \frac{1}{2}y') \\ 1 \\ (\frac{\sqrt{3}}{2}x' - \frac{1}{2}y')(\frac{1}{2}x' + \frac{\sqrt{3}}{2}y')z' \end{bmatrix} = \begin{bmatrix} \frac{3}{4}x'z' - \frac{\sqrt{3}}{4}z'y' + \frac{1}{2} \\ -\frac{\sqrt{3}}{4}x'z' + \frac{1}{4}y'z' + \frac{\sqrt{3}}{2} \\ \frac{\sqrt{3}}{4}z'(x'^2 - y'^2) + \frac{1}{2}x'y'z' \end{bmatrix}.$$

(c) Because $\nabla f = yz\mathbf{i} + xz\mathbf{j} + xy\mathbf{k}$, we can substitute $x = x(x',y',z')$, $y = y(x',y',z')$ and $z = z(x',y',z')$ into the gradient and rotate into the new basis by

$$\nabla f'(x',y',z') = \begin{bmatrix} \frac{\sqrt{3}}{2} & \frac{1}{2} & 0 \\ -\frac{1}{2} & \frac{\sqrt{3}}{2} & 0 \\ 0 & 0 & 1 \end{bmatrix} \begin{bmatrix} z'(\frac{1}{2}x' + \frac{\sqrt{3}}{2}y') \\ z'(\frac{\sqrt{3}}{2}x' - \frac{1}{2}y') \\ (\frac{\sqrt{3}}{2}x' - \frac{1}{2}y')(\frac{1}{2}x' + \frac{\sqrt{3}}{2}y') \end{bmatrix} = \begin{bmatrix} \frac{3}{2}x'z' + \frac{1}{2}y'z' \\ \frac{1}{2}x'z' - \frac{\sqrt{3}}{2}y'z' \\ \frac{\sqrt{3}}{4}(x'^2 - y'^2) + \frac{1}{2}x'y' \end{bmatrix}.$$

5.8. LINEAR ORTHOGONAL TRANSFORMATIONS

(d) The divergence in the new coordinates is $\nabla' \mathbf{V}' = \frac{\partial V'_1}{\partial x'} + \frac{\partial V'_2}{\partial y'} + \frac{\partial V'_3}{\partial z'} = \frac{3}{4}z' + \frac{1}{4}z' + \frac{\sqrt{3}}{4}(x'^2 - y'^2) + \frac{1}{2}x'y'$

(e) We can do this either by finding the curl first then converting the vector field, or by converting the vector field first then finding the curl. We will do the former. The curl of $\mathbf{V}(x,y,z) = xz\mathbf{i} + \mathbf{j} + xyz\mathbf{k}$ is $(xz)\mathbf{i} + (x - yz)\mathbf{j}$, and using the conversion $\begin{bmatrix} x \\ y \\ z \end{bmatrix} = \begin{bmatrix} \cos\frac{\pi}{6} & -\sin\frac{\pi}{6} & 0 \\ \sin\frac{\pi}{6} & \cos\frac{\pi}{6} & 0 \\ 0 & 0 & 1 \end{bmatrix} \begin{bmatrix} x' \\ y' \\ z' \end{bmatrix} = \begin{bmatrix} \frac{\sqrt{3}}{2}x' - \frac{1}{2}y' \\ \frac{1}{2}x' + \frac{\sqrt{3}}{2}y' \\ z' \end{bmatrix}$, we compute $\nabla' \times \mathbf{V}'(x', y', z') =$

$= \begin{bmatrix} \cos\frac{\pi}{6} & \sin\frac{\pi}{6} & 0 \\ -\sin\frac{\pi}{6} & \cos\frac{\pi}{6} & 0 \\ 0 & 0 & 1 \end{bmatrix} \begin{bmatrix} \left(\frac{\sqrt{3}}{2}x' - \frac{1}{2}y'\right)z' \\ \frac{\sqrt{3}}{2}x' - \frac{1}{2}y' - \left(\frac{1}{2}x' + \frac{\sqrt{3}}{2}y'\right)z' \\ 0 \end{bmatrix} =$

$= \begin{bmatrix} \frac{2\sqrt{3}-1}{8}x' - \frac{\sqrt{3}}{8}y' + \frac{4-\sqrt{3}}{8}x'z' - \frac{3+2\sqrt{3}}{8}y'z' \\ \frac{6-\sqrt{3}}{8}x' - \frac{3}{8}y' + \frac{1-4\sqrt{3}}{8}x'z' + \frac{\sqrt{3}-6}{8}z'y' \\ 0 \end{bmatrix}$

7. Verify that for the vector field in eq. (5.117), the computation of **curlV** via eq. (5.123) in the old system leads to the same vector field as the computation of **curlV** via eq. (5.124) in the new system, for the transformation of example 5.6.

   Solution: Equation (5.117) is $\mathbf{V} = \mathbf{i} + (yz)\mathbf{j} + (x^2 + y^2)\mathbf{k}$, so $\nabla \times \mathbf{V} = y\mathbf{i} - 2x\mathbf{j}$. For rotation about the z axis by $\pi/6$, $x = \frac{\sqrt{3}}{2}x' - \frac{1}{2}y'$, $y = \frac{1}{2}x' + \frac{\sqrt{3}}{2}y'$ and $z = z'$, so we compute $\mathbf{V}'(x', y', z') =$

$\begin{bmatrix} \frac{\sqrt{3}}{2} & \frac{1}{2} & 0 \\ -\frac{1}{2} & \frac{\sqrt{3}}{2} & 0 \\ 0 & 0 & 1 \end{bmatrix} \begin{bmatrix} 1 \\ (\frac{1}{2}x' + \frac{\sqrt{3}}{2}y')z' \\ (\frac{\sqrt{3}}{2}x' - \frac{1}{2}y')^2 + (\frac{1}{2}x' + \frac{\sqrt{3}}{2}y')^2 \end{bmatrix} = \begin{bmatrix} \frac{\sqrt{3}}{2} & \frac{1}{2} & 0 \\ -\frac{1}{2} & \frac{\sqrt{3}}{2} & 0 \\ 0 & 0 & 1 \end{bmatrix} \begin{bmatrix} 1 \\ \frac{1}{2}x'z' + \frac{\sqrt{3}}{2}y'z' \\ x'^2 + y'^2 \end{bmatrix} =$

$\begin{bmatrix} \frac{\sqrt{3}}{2} + \frac{1}{4}x'z' + \frac{\sqrt{3}}{4}y'z' \\ -\frac{1}{2} + \frac{\sqrt{3}}{4}x'z' + \frac{3}{4}y'z' \\ x'^2 + y'^2 \end{bmatrix}$. The curl in the new coordinates is

$\nabla' \times \mathbf{V}' = \begin{vmatrix} \mathbf{i}' & \mathbf{j}' & \mathbf{k}' \\ \partial_{x'} & \partial_{y'} & \partial_{z'} \\ \frac{\sqrt{3}}{2} + \frac{1}{4}x'z' + \frac{\sqrt{3}}{4}y'z' & -\frac{1}{2} + \frac{\sqrt{3}}{4}x'z' + \frac{3}{4}y'z' & x'^2 + y'^2 \end{vmatrix} =$

$= \left(\frac{5}{4}y' - \frac{\sqrt{3}}{4}x'\right)\mathbf{i}' + \left(\frac{\sqrt{3}}{4}y' - \frac{7}{4}x'\right)\mathbf{j}'.$

Beginning instead with $\nabla \times \mathbf{V} = y\mathbf{i} - 2x\mathbf{j}$, we compute

$\nabla' \times \mathbf{V}' = \begin{bmatrix} \frac{\sqrt{3}}{2} & \frac{1}{2} & 0 \\ -\frac{1}{2} & \frac{\sqrt{3}}{2} & 0 \\ 0 & 0 & 1 \end{bmatrix} \begin{bmatrix} \frac{1}{2}x' + \frac{\sqrt{3}}{2}y' \\ -2(\frac{\sqrt{3}}{2}x' - \frac{1}{2}y') \\ 0 \end{bmatrix} = \left(\frac{5}{4}y' - \frac{\sqrt{3}}{4}x'\right)\mathbf{i}' + \left(\frac{\sqrt{3}}{4}y' - \frac{7}{4}x'\right)\mathbf{j}'$

9. Repeat exercise 8 for the orthogonal linear transformation plus shift of eq. (5.129). *Note:* Exercise 8 states: Modeling example 5.11, give a direct proof of the "invariance" under a general linear orthogonal transformation of eq. (5.102), of

   (a) the divergence of an arbitrary vector field.
   (b) the curl of an arbitrary vector field.

   Solution:

   (a) Equation (5.129) gives an affine transformation $\begin{bmatrix} x \\ y \\ z \end{bmatrix} = \begin{bmatrix} x_0 \\ y_0 \\ z_0 \end{bmatrix} + J \begin{bmatrix} x' \\ y' \\ z' \end{bmatrix}$.

   The divergence of an arbitrary vector field can be written

5.8. LINEAR ORTHOGONAL TRANSFORMATIONS

$$\nabla' \cdot \mathbf{V}' = \left(\frac{\partial}{\partial x'}\mathbf{i}' + \frac{\partial}{\partial y'}\mathbf{j}' + \frac{\partial}{\partial z'}\mathbf{k}'\right) \cdot (V_1\mathbf{i}' + V_2\mathbf{j}' + V_3\mathbf{k}') = \frac{\partial V_1(x(x',y',z'),y(x',y',z'),z(x',y',z'))}{\partial x'}$$
$$+ \frac{\partial V_2(x(x',y',z'),y(x',y',z'),z(x',y',z'))}{\partial y'} + \frac{\partial V_3(x(x',y',z'),y(x',y',z'),z(x',y',z'))}{\partial z'}$$
$$= \left(\frac{\partial V_1}{\partial x}\frac{\partial x}{\partial x'} + \frac{\partial V_1}{\partial y}\frac{\partial y}{\partial x'} + \frac{\partial V_1}{\partial z}\frac{\partial z}{\partial x'}\right) + \left(\frac{\partial V_2}{\partial x}\frac{\partial x}{\partial x'} + \frac{\partial V_2}{\partial y}\frac{\partial y}{\partial x'} + \frac{\partial V_2}{\partial z}\frac{\partial z}{\partial x'}\right) + \left(\frac{\partial V_3}{\partial x}\frac{\partial x}{\partial x'} + \frac{\partial V_3}{\partial y}\frac{\partial y}{\partial x'} + \frac{\partial V_3}{\partial z}\frac{\partial z}{\partial x'}\right).$$

Using equations (5.102)

$$\frac{\partial x}{\partial x'} = \mathbf{i}' \cdot \mathbf{i} \quad \frac{\partial x}{\partial y'} = \mathbf{j}' \cdot \mathbf{i} \quad \frac{\partial x}{\partial z'} = \mathbf{k}' \cdot \mathbf{i}$$
$$\frac{\partial y}{\partial x'} = \mathbf{i}' \cdot \mathbf{j} \quad \frac{\partial y}{\partial y'} = \mathbf{j}' \cdot \mathbf{j} \quad \frac{\partial y}{\partial z'} = \mathbf{k}' \cdot \mathbf{j} \quad \text{or } J = \begin{bmatrix} \mathbf{i}' \cdot \mathbf{i} & \mathbf{j}' \cdot \mathbf{i} & \mathbf{k}' \cdot \mathbf{i} \\ \mathbf{i}' \cdot \mathbf{j} & \mathbf{j}' \cdot \mathbf{j} & \mathbf{k}' \cdot \mathbf{j} \\ \mathbf{i}' \cdot \mathbf{k} & \mathbf{j}' \cdot \mathbf{k} & \mathbf{k}' \cdot \mathbf{k} \end{bmatrix} = \begin{bmatrix} \frac{\partial x}{\partial x'} & \frac{\partial x}{\partial y'} & \frac{\partial x}{\partial z'} \\ \frac{\partial y}{\partial x'} & \frac{\partial y}{\partial y'} & \frac{\partial y}{\partial z'} \\ \frac{\partial z}{\partial x'} & \frac{\partial z}{\partial y'} & \frac{\partial z}{\partial z'} \end{bmatrix},$$
$$\frac{\partial z}{\partial x'} = \mathbf{i}' \cdot \mathbf{k} \quad \frac{\partial z}{\partial y'} = \mathbf{j}' \cdot \mathbf{k} \quad \frac{\partial z}{\partial z'} = \mathbf{k}' \cdot \mathbf{k}$$

we can see that this is identical to

$$\nabla \cdot \mathbf{V} = \nabla \cdot (J\mathbf{V}') = \left[\frac{\partial}{\partial x}\mathbf{i} + \frac{\partial}{\partial y}\mathbf{j} + \frac{\partial}{\partial z}\mathbf{k}\right] J \begin{bmatrix} V_1(x(x',y',z'),y(x',y',z'),z(x',y',z')) \\ V_2(x(x',y',z'),y(x',y',z'),z(x',y',z')) \\ V_3(x(x',y',z'),y(x',y',z'),z(x',y',z')) \end{bmatrix}$$

$$= \left[\frac{\partial}{\partial x}\mathbf{i} + \frac{\partial}{\partial y}\mathbf{j} + \frac{\partial}{\partial z}\mathbf{k}\right] \begin{bmatrix} \frac{\partial x}{\partial x'} & \frac{\partial x}{\partial y'} & \frac{\partial x}{\partial z'} \\ \frac{\partial y}{\partial x'} & \frac{\partial y}{\partial y'} & \frac{\partial y}{\partial z'} \\ \frac{\partial z}{\partial x'} & \frac{\partial z}{\partial y'} & \frac{\partial z}{\partial z'} \end{bmatrix} \begin{bmatrix} V_1(x(x',y',z'),y(x',y',z'),z(x',y',z')) \\ V_2(x(x',y',z'),y(x',y',z'),z(x',y',z')) \\ V_3(x(x',y',z'),y(x',y',z'),z(x',y',z')) \end{bmatrix}$$

$$= \left[\frac{\partial}{\partial x}\mathbf{i} + \frac{\partial}{\partial y}\mathbf{j} + \frac{\partial}{\partial z}\mathbf{k}\right] \cdot$$
$$\left[\left(V_1\frac{\partial x}{\partial x'} + V_2\frac{\partial x}{\partial y'} + V_3\frac{\partial x}{\partial z'}\right)\mathbf{i} + \left(V_1\frac{\partial y}{\partial x'} + V_2\frac{\partial y}{\partial y'} + V_3\frac{\partial y}{\partial z'}\right)\mathbf{j} + \left(V_1\frac{\partial z}{\partial x'} + V_2\frac{\partial z}{\partial y'} + V_3\frac{\partial z}{\partial z'}\right)\mathbf{k}\right], \text{ or}$$
$$\left(\frac{\partial V_1}{\partial x}\frac{\partial x}{\partial x'} + \frac{\partial V_2}{\partial x}\frac{\partial x}{\partial y'} + \frac{\partial V_3}{\partial x}\frac{\partial x}{\partial z'}\right) + \left(\frac{\partial V_1}{\partial y}\frac{\partial y}{\partial x'} + \frac{\partial V_2}{\partial y}\frac{\partial y}{\partial y'} + \frac{\partial V_3}{\partial y}\frac{\partial y}{\partial z'}\right) + \left(\frac{\partial V_1}{\partial z}\frac{\partial z}{\partial x'} + \frac{\partial V_2}{\partial z}\frac{\partial z}{\partial y'} + \frac{\partial V_3}{\partial z}\frac{\partial z}{\partial z'}\right),$$

which is just
$$\left(\frac{\partial V_1}{\partial x}\frac{\partial x}{\partial x'} + \frac{\partial V_1}{\partial y}\frac{\partial y}{\partial x'} + \frac{\partial V_1}{\partial z}\frac{\partial z}{\partial x'}\right) + \left(\frac{\partial V_2}{\partial x}\frac{\partial x}{\partial y'} + \frac{\partial V_2}{\partial y}\frac{\partial y}{\partial y'} + \frac{\partial V_2}{\partial z}\frac{\partial z}{\partial y'}\right) + \left(\frac{\partial V_3}{\partial x}\frac{\partial x}{\partial z'} + \frac{\partial V_3}{\partial y}\frac{\partial y}{\partial z'} + \frac{\partial V_3}{\partial z}\frac{\partial z}{\partial z'}\right)$$

rearranged.

To modify the argument for the affine transformation, just write

$$\begin{bmatrix} \bar{x} \\ \bar{y} \\ \bar{z} \end{bmatrix} = \begin{bmatrix} x - x_0 \\ y - y_0 \\ z - z_0 \end{bmatrix} = J \begin{bmatrix} x' \\ y' \\ z' \end{bmatrix}$$

and substitute $\bar{x}$, $\bar{y}$ and $\bar{z}$ for x, y and z everywhere in the argument, or note that $V_1 = V_1(x(x_0, x', y', z'), y(y_0, x', y', z'), z(z_0, x', y', z'))$ and likewise for V_2 and V_3. Because x_0, y_0 and z_0 are constants, the partial derivatives with respect to these variables are all zero, and the argument goes through as before.

(b) The curl in a rotated frame can be written $\nabla' \times \mathbf{V}' = \left(\frac{\partial V_3}{\partial y'} - \frac{\partial V_2}{\partial z'}\right)\mathbf{i}' + \left(\frac{\partial V_1}{\partial z'} - \frac{\partial V_3}{\partial x'}\right)\mathbf{j}' + \left(\frac{\partial V_2}{\partial x'} - \frac{\partial V_1}{\partial y'}\right)\mathbf{k}'$. Expanding this, we get,

$$\nabla' \times \mathbf{V}' =$$
$$\left[\left(\frac{\partial V_3}{\partial x}\frac{\partial x}{\partial y'} + \frac{\partial V_3}{\partial y}\frac{\partial y}{\partial y'} + \frac{\partial V_3}{\partial z}\frac{\partial z}{\partial y'}\right) - \left(\frac{\partial V_2}{\partial x}\frac{\partial x}{\partial z'} + \frac{\partial V_2}{\partial y}\frac{\partial y}{\partial z'} + \frac{\partial V_2}{\partial z}\frac{\partial z}{\partial z'}\right)\right]\mathbf{i}' +$$
$$\left[\left(\frac{\partial V_1}{\partial x}\frac{\partial x}{\partial z'} + \frac{\partial V_1}{\partial y}\frac{\partial y}{\partial z'} + \frac{\partial V_1}{\partial z}\frac{\partial z}{\partial z'}\right) - \left(\frac{\partial V_3}{\partial x}\frac{\partial x}{\partial x'} + \frac{\partial V_3}{\partial y}\frac{\partial y}{\partial x'} + \frac{\partial V_3}{\partial z}\frac{\partial z}{\partial x'}\right)\right]\mathbf{j}' +$$
$$\left[\left(\frac{\partial V_2}{\partial x}\frac{\partial x}{\partial x'} + \frac{\partial V_2}{\partial y}\frac{\partial y}{\partial x'} + \frac{\partial V_2}{\partial z}\frac{\partial z}{\partial x'}\right) - \left(\frac{\partial V_1}{\partial x}\frac{\partial x}{\partial y'} + \frac{\partial V_1}{\partial y}\frac{\partial y}{\partial y'} + \frac{\partial V_1}{\partial z}\frac{\partial z}{\partial y'}\right)\right]\mathbf{k}'.$$

Letting $\mathbf{J}_1$, $\mathbf{J}_2$ and $\mathbf{J}_3$ stand for the first, second and third columns of J respectively, we can write this as $(\mathbf{J}_2 \cdot \nabla V_3 - \mathbf{J}_3 \cdot \nabla V_2)\mathbf{i}' + (\mathbf{J}_3 \cdot \nabla V_1 - \mathbf{J}_1 \cdot \nabla V_3)\mathbf{j}' + (\mathbf{J}_1 \cdot \nabla V_2 - \mathbf{J}_2 \cdot \nabla V_1)\mathbf{k}'$. Take the curl and write it in the new coordinates by $\nabla' \times \mathbf{V}' = J^{-1}\begin{bmatrix} \partial_y V_3 - \partial_z V_2 \\ \partial_z V_1 - \partial_x V_3 \\ \partial_x V_2 - \partial_y V_1 \end{bmatrix}$ where the derivatives are taken with respect to the original variables x, y and z and then the functions of $x = x(x', y', z')$ and so on are substituted. Then by the chain rule,

$$J^{-1}(\nabla \times \mathbf{V}) =$$
$$J^{-1} \begin{bmatrix} \left(\frac{\partial V_3}{\partial x'}\frac{\partial x'}{\partial y} + \frac{\partial V_3}{\partial y'}\frac{\partial y'}{\partial y} + \frac{\partial V_3}{\partial z'}\frac{\partial z'}{\partial y}\right) - \left(\frac{\partial V_2}{\partial x'}\frac{\partial x'}{\partial z} + \frac{\partial V_2}{\partial y'}\frac{\partial y'}{\partial z} + \frac{\partial V_2}{\partial z'}\frac{\partial z'}{\partial z}\right) \\ \left(\frac{\partial V_1}{\partial x'}\frac{\partial x'}{\partial z} + \frac{\partial V_1}{\partial y'}\frac{\partial y'}{\partial z} + \frac{\partial V_1}{\partial z'}\frac{\partial z'}{\partial z}\right) - \left(\frac{\partial V_3}{\partial x'}\frac{\partial x'}{\partial x} + \frac{\partial V_3}{\partial y'}\frac{\partial y'}{\partial x} + \frac{\partial V_3}{\partial z'}\frac{\partial z'}{\partial x}\right) \\ \left(\frac{\partial V_2}{\partial x'}\frac{\partial x'}{\partial x} + \frac{\partial V_2}{\partial y'}\frac{\partial y'}{\partial x} + \frac{\partial V_2}{\partial z'}\frac{\partial z'}{\partial x}\right) - \left(\frac{\partial V_1}{\partial x'}\frac{\partial x'}{\partial y} + \frac{\partial V_1}{\partial y'}\frac{\partial y'}{\partial y} + \frac{\partial V_1}{\partial z'}\frac{\partial z'}{\partial y}\right) \end{bmatrix}.$$

5.8. LINEAR ORTHOGONAL TRANSFORMATIONS

11. What is the element of arc length, $ds = (dx^2 + dy^2 + dz^2)^{1/2}$ in the new coordinate system?
Solution: Because the coordinate system is orthogonal and the coordinate curves are straight lines, the scale factors are all equal to one, so $ds' = (dx'^2 + dy'^2 + dz'^2)^{1/2}$.

13. Suppose a new x', y', z' coordinate system is related to the original system by a linear orthogonal transformation described by a matrix J, as in eq. (5.103), and a "still newer" x'', y'', z'' coordinate system is "related" to the "new" system by a linear orthogonal transformation generated by a matrix K. Show that the x'', y'', z'' system is related to the original x, y, z system by a linear orthogonal transformation, with the associated matrix given by JK. Prove directly that the product of two orthogonal matrices is orthogonal. What interpretation can you give to the columns of JK?
Solution: Let $\mathbf{x}$ represent the vector $x\mathbf{i} + y\mathbf{j} + z\mathbf{k}$, and similarly for $\mathbf{x}'$ and $\mathbf{x}''$. Then $\mathbf{x} = J\mathbf{x}'$, $\mathbf{x}' = K\mathbf{x}''$, and $\mathbf{x} = JK\mathbf{x}''$. Because J and K are orthogonal, all we need to prove is that JK is orthogonal. Recall that a matrix is orthogonal if its transpose is equal to its inverse, or $J^T = J^{-1}$. The prth entry of JK is $(JK)_{pr} = \sum_q J_{pq} K_{qr} = \mathbf{J}_p \cdot \mathbf{K}_r$, where $\mathbf{J}_p$ is the pth row vector of J and $\mathbf{K}_r$ is the rth column vector of K, and q indexes their entries. Because the transpose reverses the indices, $[(JK)]^T_{pr} = (\mathbf{J}_p \cdot \mathbf{K}_r)_{rp}$. But r is the index associated with $\mathbf{K}$ and p is the index associated with $\mathbf{J}$, so this is $(\mathbf{K}_r \cdot \mathbf{J}_p)_{rp} = (K^T J^T)_{rp}$, where we now interpret $\mathbf{J}_p$ as the pth column vector of J and $\mathbf{K}_r$ as the rth row vector of K. Therefore $(JK)^T = K^T J^T = K^{-1} J^{-1} = (JK)^{-1}$, and the result is proved. The columns of JK are vectors in the direction of the x'', y'' and z'' coordinate axes.

15. Show directly that scalar products are preserved under the general linear orthogonal transformation of eq. (5.110), that is, show that $V_1 W_1 + V_2 W_2 + V_3 W_3 = V'_1 W'_1 + V'_2 W'_2 + V'_3 W'_3$ when the components of $\mathbf{V}$ and $\mathbf{W}$ are related by eq. (5.110). This provides verification of the (obvious) fact that lengths and angles are preserved under these transformations.

Solution: Equation (5.110) states $\begin{bmatrix} V_1 \\ V_2 \\ V_3 \end{bmatrix} = J \begin{bmatrix} V'_1 \\ V'_2 \\ V'_3 \end{bmatrix}$ and $\begin{bmatrix} V'_1 \\ V'_2 \\ V'_3 \end{bmatrix} = J^{-1} \begin{bmatrix} V_1 \\ V_2 \\ V_3 \end{bmatrix}$, so writing the same for $\mathbf{W}$ and $\mathbf{W}'$, we have $\begin{bmatrix} W_1 \\ W_2 \\ W_3 \end{bmatrix} = J \begin{bmatrix} W'_1 \\ W'_2 \\ W'_3 \end{bmatrix}$ and $\begin{bmatrix} W'_1 \\ W'_2 \\ W'_3 \end{bmatrix} = J^{-1} \begin{bmatrix} W_1 \\ W_2 \\ W_3 \end{bmatrix}$. Then $\mathbf{V} \cdot \mathbf{W} = (J\mathbf{V}') \cdot (J\mathbf{W}') = (J\mathbf{V}')^T (J\mathbf{W}')$ (by the equivalence of the dot product and multiplication of row and column matrices). Now we have $(J\mathbf{V}')^T (J\mathbf{W}') = \mathbf{V}'^T J^T J \mathbf{W}'$, but J is orthogonal, so $J^T = J^{-1}$ and $\mathbf{V}'^T J^T J \mathbf{W}' = \mathbf{V}'^T J^{-1} J \mathbf{W}' = \mathbf{V}'^T \mathbf{W}' = \mathbf{V}' \cdot \mathbf{W}'$, again using the equivalence of the dot product and multiplication of row and column matrices.

17. Based on the last two exercises, can you derive *Euler's theorem*: every transformation of the form of eq. (5.102) can be described as a rotation of the coordinate system about some straight line through the origin?
Solution: Equation (5.102) reads
$x = x'\mathbf{i}' \cdot \mathbf{i} + y'\mathbf{j}' \cdot \mathbf{i} + z'\mathbf{k}' \cdot \mathbf{i}$
$y = x'\mathbf{i}' \cdot \mathbf{j} + y'\mathbf{j}' \cdot \mathbf{j} + z'\mathbf{k}' \cdot \mathbf{j}$
$z = x'\mathbf{i}' \cdot \mathbf{k} + y'\mathbf{j}' \cdot \mathbf{k} + z'\mathbf{k}' \cdot \mathbf{k}$,
and is equivalent to $\begin{bmatrix} x \\ y \\ z \end{bmatrix} = \begin{bmatrix} \mathbf{i}' \cdot \mathbf{i} & \mathbf{j}' \cdot \mathbf{i} & \mathbf{k}' \cdot \mathbf{i} \\ \mathbf{i}' \cdot \mathbf{j} & \mathbf{j}' \cdot \mathbf{j} & \mathbf{k}' \cdot \mathbf{j} \\ \mathbf{i}' \cdot \mathbf{k} & \mathbf{j}' \cdot \mathbf{k} & \mathbf{k}' \cdot \mathbf{k} \end{bmatrix} \begin{bmatrix} x' \\ y' \\ z' \end{bmatrix}$. Because $\begin{bmatrix} \mathbf{i} \\ \mathbf{j} \\ \mathbf{k} \end{bmatrix} = \begin{bmatrix} 1 & 0 & 0 \\ 0 & 1 & 0 \\ 0 & 0 & 1 \end{bmatrix} = [\ \mathbf{i}\ \mathbf{j}\ \mathbf{k}\ ]$, that is, the identity matrix equals its inverse equals its transpose, it is easy to see that $\begin{bmatrix} x \\ y \\ z \end{bmatrix} = [\ \mathbf{i}\ \mathbf{j}\ \mathbf{k}\ ] \begin{bmatrix} x \\ y \\ z \end{bmatrix} = [\ \mathbf{i}'\ \mathbf{j}'\ \mathbf{k}'\ ] \begin{bmatrix} x' \\ y' \\ z' \end{bmatrix}$ implies $\begin{bmatrix} x \\ y \\ z \end{bmatrix} = \begin{bmatrix} \mathbf{i} \\ \mathbf{j} \\ \mathbf{k} \end{bmatrix} [\ \mathbf{i}'\ \mathbf{j}'\ \mathbf{k}'\ ] \begin{bmatrix} x' \\ y' \\ z' \end{bmatrix} = \begin{bmatrix} \mathbf{i}' \cdot \mathbf{i} & \mathbf{j}' \cdot \mathbf{i} & \mathbf{k}' \cdot \mathbf{i} \\ \mathbf{i}' \cdot \mathbf{j} & \mathbf{j}' \cdot \mathbf{j} & \mathbf{k}' \cdot \mathbf{j} \\ \mathbf{i}' \cdot \mathbf{k} & \mathbf{j}' \cdot \mathbf{k} & \mathbf{k}' \cdot \mathbf{k} \end{bmatrix} \begin{bmatrix} x' \\ y' \\ z' \end{bmatrix}$. Thus, transformations of the given form can be rewritten to give $x\mathbf{i} + y\mathbf{j} + z\mathbf{k}$ in terms of the coefficients $x'\mathbf{i}' + y'\mathbf{j}' + z'\mathbf{k}'$ of a rotated coordinate system.
A reasonable constructive but non-rigorous argument that we can always find a transformation that rotates about any line through the origin can be made this way: we can always find a vector, say $\mathbf{v}$ in the direction of any straight line L through the origin, and by a sequence of at most two rotations can rotate it into one of the coordinate axes. We can then rotate about this axis, and by a sequence of at most two rotations, rotate $\mathbf{v}$ back into its original position.

19. As a partial converse to the theory developed in this section, suppose we *begin* with a transformation

5.8. LINEAR ORTHOGONAL TRANSFORMATIONS

of coordinates defined by $\begin{bmatrix} x \\ y \\ z \end{bmatrix} = J \begin{bmatrix} x' \\ y' \\ z' \end{bmatrix}$ and we know only that J is an orthogonal matrix.

(a) By examining the points with new coordinates $(1,0,0)$, $(0,1,0)$ and $(0,0,1)$ in turn, show that the columns of J, interpreted as vectors expressed in the old system, point along the new x', y', and z' axes.

(b) Exploit the orthogonality of J to prove that the new axes are mutually orthogonal and that $(x'^2 + y'^2 + z'^2)^{1/2}$ equals the distance of (x', y', z') from the origin.

(c) From (a) and (b) we may conclude that the new system is a bona fide cartesian coordinate system. However, it may be left-handed, as the simple example $\begin{bmatrix} x \\ y \\ z \end{bmatrix} = \begin{bmatrix} 1 & 0 & 0 \\ 0 & 1 & 0 \\ 0 & 0 & -1 \end{bmatrix} \begin{bmatrix} x' \\ y' \\ z' \end{bmatrix}$ shows. What modification of rules in eqs. (5.110), (5.112), (5.116), (5.120), (5.122), and (5.124) have to be made in transforming to a left handed system?

(d) How can you determine, from the matrix J, whether or not the new system is right-handed?

Solution:

(a) Let J have columns $\mathbf{J}_1$, $\mathbf{J}_2$ and $\mathbf{J}_3$. A vector $\mathbf{X} = \begin{bmatrix} x \\ y \\ z \end{bmatrix} = [\,\mathbf{J}_1\ \mathbf{J}_2\ \mathbf{J}_3\,] \begin{bmatrix} x' \\ y' \\ z' \end{bmatrix} = x'\mathbf{J}_1 + y'\mathbf{J}_2 + z'\mathbf{J}_3$, so $\mathbf{J}_1$, $\mathbf{J}_2$ and $\mathbf{J}_3$ are vectors in the direction of the coordinate x', y' and z' axes. More specifically, take the points $\begin{bmatrix} 1 \\ 0 \\ 0 \end{bmatrix}$, $\begin{bmatrix} 0 \\ 1 \\ 0 \end{bmatrix}$ and $\begin{bmatrix} 0 \\ 0 \\ 1 \end{bmatrix}$ and multiply these by J in turns:

$[\,\mathbf{J}_1\ \mathbf{J}_2\ \mathbf{J}_3\,] \begin{bmatrix} 1 \\ 0 \\ 0 \end{bmatrix} = \mathbf{J}_1$, $[\,\mathbf{J}_1\ \mathbf{J}_2\ \mathbf{J}_3\,] \begin{bmatrix} 0 \\ 1 \\ 0 \end{bmatrix} = \mathbf{J}_2$ and $[\,\mathbf{J}_1\ \mathbf{J}_2\ \mathbf{J}_3\,] \begin{bmatrix} 0 \\ 0 \\ 1 \end{bmatrix} = \mathbf{J}_3$, so again, it is easy to see that the columns of J are the new coordinate axes.

(b) Because J is orthogonal, $J^T = J^{-1}$ so $JJ^{-1} = JJ^T = I$, and thus $\mathbf{J}_1 \cdot \mathbf{J}_1 = \mathbf{J}_2 \cdot \mathbf{J}_2 = \mathbf{J}_3 \cdot \mathbf{J}_3 = 1$ and $\mathbf{J}_i \cdot \mathbf{J}_j = 0$ for $i \neq j$. The distance of (x', y', z') from the origin is the same as the distance of (x, y, z) from the origin, that is,

$$(x^2 + y^2 + z^2)^{1/2} = ((x\mathbf{i} + y\mathbf{j} + z\mathbf{k}) \cdot (x\mathbf{i} + y\mathbf{j} + z\mathbf{k}))^{1/2} = \left([x,y,z] \begin{bmatrix} x \\ y \\ z \end{bmatrix}\right)^{1/2}$$

$$= \left([x,y,z]I^T I \begin{bmatrix} x \\ y \\ z \end{bmatrix}\right)^{1/2} = \left([x',y',z']J^T J \begin{bmatrix} x' \\ y' \\ z' \end{bmatrix}\right)^{1/2} = \left([x',y',z'] \begin{bmatrix} x' \\ y' \\ z' \end{bmatrix}\right)^{1/2}$$

$$= (x'^2 + y'^2 + z'^2)^{1/2}.$$

(c) When the coordinate system is left-handed, the signed volume element $\frac{\partial \mathbf{R}}{\partial u_1} \cdot \left(\frac{\partial \mathbf{R}}{\partial u_2} \times \frac{\partial \mathbf{R}}{\partial u_3}\right) du_1 du_2 du_3$ has the opposite sign, as does the curl and the cross product. The rotation matrices must be taken to rotate in the *clockwise* direction if their signs are not changed. With these changes the change of coordinates formulas will yield valid results.

(d) Take the determinant. The determinant of a left handed orthogonal system is -1, while that of a right handed system is 1.

21. Derive the matrix for the transformation generated by rotating the x, y, and z system through an angle $\pi/2$ about the straight line through the origin parallel to $\mathbf{i} + \mathbf{j} + \mathbf{k}$.

Solution: In order to affect a rotation about an arbitrary vector, we rotate the vector into a coordinate axis, apply a rotation about that axis, then rotate the vector back to its original position. There are a number of ways to do this for this problem, so here is one: Rotate $\mathbf{i} + \mathbf{j} + b\mathbf{k}$ clockwise about the z axis by the matrix $A = \begin{bmatrix} \cos \pi/2 & \sin \pi/2 & 0 \\ -\sin \pi/2 & \cos \pi/2 & 0 \\ 0 & 0 & 1 \end{bmatrix}$ so that it sits above the x axis. Now lower it into the x axis by rotating counterclockwise about the y axis using the matrix $B = \begin{bmatrix} \cos \gamma & 0 & \sin \gamma \\ 0 & 1 & 0 \\ \cos \gamma & 0 & -\sin \gamma \end{bmatrix}$. The values of γ are found by noting that the vector $\begin{bmatrix} 1 \\ 1 \\ 1 \end{bmatrix}$ points to the corner of a cube from the diagonally opposite corner at the origin, and that its length is $\sqrt{3}$, its projection onto the xy plane has length $\sqrt{2}$ and it's projection onto an edge of the cube parallel to the z axis has length 1. Thus $\sqrt{3}\cos\gamma = \sqrt{2}$ or $\cos\gamma = \sqrt{\frac{2}{3}}$ and $\sqrt{3}\sin\gamma = 1$ or $\sin\gamma = \sqrt{\frac{1}{3}}$. The matrix thus has the specific form $B = \begin{bmatrix} \sqrt{\frac{2}{3}} & 0 & \sqrt{\frac{1}{3}} \\ 0 & 1 & 0 \\ \sqrt{\frac{2}{3}} & 0 & -\sqrt{\frac{1}{3}} \end{bmatrix}$. Now to rotate by $\pi/2$ about the x axis we use the matrix $R = \begin{bmatrix} 1 & 0 & 0 \\ 0 & \cos\pi/2 & -\sin\pi/2 \\ 0 & \sin\pi/2 & \cos\pi/2 \end{bmatrix}$. The transformation so far is $RBA\mathbf{X}$, but in order to rotate about $\mathbf{X} = \begin{bmatrix} 1 \\ 1 \\ 1 \end{bmatrix}$ we need to move the transformed vector back to its original position, that is, $\mathbf{X} = A^{-1}B^{-1}RBA\mathbf{X}$. If we have done this right, any vector in the direction of $\mathbf{X} = \begin{bmatrix} 1 \\ 1 \\ 1 \end{bmatrix}$ will be invariant under $\mathbf{X} = A^{-1}B^{-1}RBA$, because we are rotating about that vector.

Specifically, we get $A^{-1}B^{-1}RBA = \begin{bmatrix} \frac{1}{3} & \frac{1}{3}-\frac{\sqrt{3}}{3} & \frac{1}{3}+\frac{\sqrt{3}}{3} \\ \frac{1}{3}+\frac{\sqrt{3}}{3} & \frac{1}{3} & \frac{1}{3}-\frac{\sqrt{3}}{3} \\ \frac{1}{3}-\frac{\sqrt{3}}{3} & \frac{1}{3}+\frac{\sqrt{3}}{3} & \frac{1}{3} \end{bmatrix}$, and the result of multiplying this times the vector $\mathbf{i}+\mathbf{j}+\mathbf{k}$ is the same as adding the columns — we can see that this gives us $\mathbf{i}+\mathbf{j}+\mathbf{k}$ back.

23. (*Significance of the Jacobian*) Obviously the transformation equations (3.69) for orthogonal coordinates are, in general, nonlinear. However, the relation $d\mathbf{R} = \frac{\partial \mathbf{R}}{\partial u_1}du_1 + \frac{\partial \mathbf{R}}{\partial u_2}du_2 + \frac{\partial \mathbf{R}}{\partial u_3}du_3$ between the differentials can be viewed as a "local linearization" of eq. (3.69).

   (a) Show that eq. (5.131) can be expressed

   $$\begin{bmatrix} dx \\ dy \\ dz \end{bmatrix} = \begin{bmatrix} \frac{\partial x}{\partial u_1} & \frac{\partial x}{\partial u_2} & \frac{\partial x}{\partial u_3} \\ \frac{\partial y}{\partial u_1} & \frac{\partial y}{\partial u_2} & \frac{\partial y}{\partial u_3} \\ \frac{\partial z}{\partial u_1} & \frac{\partial z}{\partial u_2} & \frac{\partial z}{\partial u_3} \end{bmatrix} \begin{bmatrix} du_1 \\ du_2 \\ du_3 \end{bmatrix}$$

   Recall that the matrix of partial derivatives in eq. (5.132) was identified in this section as the Jacobian of the transformation in eq. (3.69). It was abbreviated $J = \frac{\partial(x,y,z)}{\partial(u_1,u_2,u_3)}$

   (b) Show that the chain rule implies that the inverse of the Jacobian is

   $$J^{-1} = \frac{\partial(u_1, u_2, u_3)}{\partial(x, y, z)} = \begin{bmatrix} \frac{\partial u_1}{\partial x} & \frac{\partial u_1}{\partial y} & \frac{\partial u_1}{\partial z} \\ \frac{\partial u_2}{\partial x} & \frac{\partial u_2}{\partial y} & \frac{\partial u_2}{\partial z} \\ \frac{\partial u_3}{\partial x} & \frac{\partial u_3}{\partial y} & \frac{\partial u_3}{\partial z} \end{bmatrix}$$

   (c) Show that the requirement of (u_1, u_2, u_3) forming orthogonal curvilinear coordinates forces the rows of J to be orthogonal. Nonetheless, J is not an orthogonal matrix. Why?
   (d) Show that the determinant of J is $h_1h_2h_3$, the factor appearing in the volume element of eq. (3.81) (should be 3.82). This prompts the mnemonic $dxdydz = \left|\frac{\partial(x,y,z)}{\partial(u_1,u_2,u_3)}\right|du_1du_2du_3$.

Solution:

(a) Equation (5.131) reads $d\mathbf{R} = \frac{\partial \mathbf{R}}{\partial u_1}du_1 + \frac{\partial \mathbf{R}}{\partial u_2}du_2 + \frac{\partial \mathbf{R}}{\partial u_3}du_3$. The components of $d\mathbf{R}$ are dx, dy and dz, and taking the dot product of the rows of the Jacobian matrix with the column vector on the right hand side of

$$\begin{bmatrix} dx \\ dy \\ dz \end{bmatrix} = \begin{bmatrix} \frac{\partial x}{\partial u_1} & \frac{\partial x}{\partial u_2} & \frac{\partial x}{\partial u_3} \\ \frac{\partial y}{\partial u_1} & \frac{\partial y}{\partial u_2} & \frac{\partial y}{\partial u_3} \\ \frac{\partial z}{\partial u_1} & \frac{\partial z}{\partial u_2} & \frac{\partial z}{\partial u_3} \end{bmatrix} \begin{bmatrix} du_1 \\ du_2 \\ du_3 \end{bmatrix}$$

gives each component of $d\mathbf{R}$.

5.8. LINEAR ORTHOGONAL TRANSFORMATIONS

(b) Suppose that $(x, y, z) = f(u_1, u_2, u_3)$ and $(u_1, u_2, u_3) = g(x, y, z)$ are linear orthogonal transformations, that can each be represented by a matrix of partial derivatives, that is, $\mathbf{X} = M\mathbf{U}$ and $\mathbf{U} = N\mathbf{X}$. Then $(x, y, z) = f(g(x, y, z))$, is equivalent to $\mathbf{X} = MN\mathbf{X}$, and the Jacobian derivative is $D(MN) = D(M)D(N)\mathbf{X} = JJ^{-1} = I$, by the chain rule (remember MN is to be thought of as the matrix version of composition of f and g) so

$$\begin{bmatrix} \frac{\partial u_1}{\partial x} & \frac{\partial u_1}{\partial y} & \frac{\partial u_1}{\partial z} \\ \frac{\partial u_2}{\partial x} & \frac{\partial u_2}{\partial y} & \frac{\partial u_2}{\partial z} \\ \frac{\partial u_3}{\partial x} & \frac{\partial u_3}{\partial y} & \frac{\partial u_3}{\partial z} \end{bmatrix} \text{ is the inverse of } \begin{bmatrix} \frac{\partial x}{\partial u_1} & \frac{\partial x}{\partial u_2} & \frac{\partial x}{\partial u_3} \\ \frac{\partial y}{\partial u_1} & \frac{\partial y}{\partial u_2} & \frac{\partial y}{\partial u_3} \\ \frac{\partial z}{\partial u_1} & \frac{\partial z}{\partial u_2} & \frac{\partial z}{\partial u_3} \end{bmatrix}.$$

Alternatively, we can write out the product

$$\begin{bmatrix} \frac{\partial x}{\partial u_1} & \frac{\partial x}{\partial u_2} & \frac{\partial x}{\partial u_3} \\ \frac{\partial y}{\partial u_1} & \frac{\partial y}{\partial u_2} & \frac{\partial y}{\partial u_3} \\ \frac{\partial z}{\partial u_1} & \frac{\partial z}{\partial u_2} & \frac{\partial z}{\partial u_3} \end{bmatrix} \begin{bmatrix} \frac{\partial u_1}{\partial x} & \frac{\partial u_1}{\partial y} & \frac{\partial u_1}{\partial z} \\ \frac{\partial u_2}{\partial x} & \frac{\partial u_2}{\partial y} & \frac{\partial u_2}{\partial z} \\ \frac{\partial u_3}{\partial x} & \frac{\partial u_3}{\partial y} & \frac{\partial u_3}{\partial z} \end{bmatrix} =$$

$$= \begin{bmatrix} \frac{\partial x}{\partial u_1}\frac{\partial u_1}{\partial x} + \frac{\partial x}{\partial u_2}\frac{\partial u_2}{\partial x} + \frac{\partial x}{\partial u_3}\frac{\partial u_3}{\partial x} & \frac{\partial x}{\partial u_1}\frac{\partial u_1}{\partial y} + \frac{\partial x}{\partial u_2}\frac{\partial u_2}{\partial y} + \frac{\partial x}{\partial u_3}\frac{\partial u_3}{\partial y} & \frac{\partial x}{\partial u_1}\frac{\partial u_1}{\partial z} + \frac{\partial x}{\partial u_2}\frac{\partial u_2}{\partial z} + \frac{\partial x}{\partial u_3}\frac{\partial u_3}{\partial z} \\ \frac{\partial y}{\partial u_1}\frac{\partial u_1}{\partial x} + \frac{\partial y}{\partial u_2}\frac{\partial u_2}{\partial x} + \frac{\partial y}{\partial u_3}\frac{\partial u_3}{\partial x} & \frac{\partial y}{\partial u_1}\frac{\partial u_1}{\partial y} + \frac{\partial y}{\partial u_2}\frac{\partial u_2}{\partial y} + \frac{\partial y}{\partial u_3}\frac{\partial u_3}{\partial y} & \frac{\partial y}{\partial u_1}\frac{\partial u_1}{\partial z} + \frac{\partial y}{\partial u_2}\frac{\partial u_2}{\partial z} + \frac{\partial y}{\partial u_3}\frac{\partial u_3}{\partial z} \\ \frac{\partial z}{\partial u_1}\frac{\partial u_1}{\partial x} + \frac{\partial z}{\partial u_2}\frac{\partial u_2}{\partial x} + \frac{\partial z}{\partial u_3}\frac{\partial u_3}{\partial x} & \frac{\partial z}{\partial u_1}\frac{\partial u_1}{\partial y} + \frac{\partial z}{\partial u_2}\frac{\partial u_2}{\partial y} + \frac{\partial z}{\partial u_3}\frac{\partial u_3}{\partial y} & \frac{\partial z}{\partial u_1}\frac{\partial u_1}{\partial z} + \frac{\partial z}{\partial u_2}\frac{\partial u_2}{\partial z} + \frac{\partial z}{\partial u_3}\frac{\partial u_3}{\partial z} \end{bmatrix},$$

which, by the chain rule is $\begin{bmatrix} \frac{\partial x}{\partial x} & \frac{\partial x}{\partial y} & \frac{\partial x}{\partial z} \\ \frac{\partial y}{\partial x} & \frac{\partial y}{\partial y} & \frac{\partial y}{\partial z} \\ \frac{\partial z}{\partial x} & \frac{\partial z}{\partial y} & \frac{\partial z}{\partial z} \end{bmatrix} = \begin{bmatrix} 1 & 0 & 0 \\ 0 & 1 & 0 \\ 0 & 0 & 1 \end{bmatrix}$.

(c) The transformation might be left-handed, so the rows J_i of J may be orthogonal, that is, $J_1 \cdot J_2 = J_2 \cdot J_3 = J_3 \cdot J_1 = 0$, while the matrix itself is not orthogonal, $J^T \neq J^{-1}$.

(d) As vectors, the columns of the Jacobian matrix $J = \begin{bmatrix} \frac{\partial x}{\partial u_1} & \frac{\partial x}{\partial u_2} & \frac{\partial x}{\partial u_3} \\ \frac{\partial y}{\partial u_1} & \frac{\partial y}{\partial u_2} & \frac{\partial y}{\partial u_3} \\ \frac{\partial z}{\partial u_1} & \frac{\partial z}{\partial u_2} & \frac{\partial z}{\partial u_3} \end{bmatrix}$ can be written $\frac{\partial \mathbf{R}}{\partial u_1} = \frac{\partial x}{\partial u_1}\mathbf{i} + \frac{\partial y}{\partial u_1}\mathbf{j} + \frac{\partial z}{\partial u_1}\mathbf{k}$, $\frac{\partial \mathbf{R}}{\partial u_2} = \frac{\partial x}{\partial u_2}\mathbf{i} + \frac{\partial y}{\partial u_2}\mathbf{j} + \frac{\partial z}{\partial u_2}\mathbf{k}$ and $\frac{\partial \mathbf{R}}{\partial u_3} = \frac{\partial x}{\partial u_3}\mathbf{i} + \frac{\partial y}{\partial u_3}\mathbf{j} + \frac{\partial z}{\partial u_3}\mathbf{k}$. These are the vectors giving the change in the position vector $\mathbf{R} = x\mathbf{i} + y\mathbf{j} + z\mathbf{k}$ along the new coordinate directions u_1, u_2 and u_3, and are thus tangent to the u_1, u_2 and u_3 coordinate curves. Just as the cross product $\frac{\partial \mathbf{R}}{\partial u} \times \frac{\partial \mathbf{R}}{\partial v}$ gives the scale of the area element $\frac{\partial \mathbf{R}}{\partial u} \times \frac{\partial \mathbf{R}}{\partial v} dudv$ for surface integration, so the triple scalar product $\frac{\partial \mathbf{R}}{\partial u_1} \cdot \left(\frac{\partial \mathbf{R}}{\partial u_2} \times \frac{\partial \mathbf{R}}{\partial u_3} \right)$ gives the scale of the volume element $\frac{\partial \mathbf{R}}{\partial u_1} \cdot \left(\frac{\partial \mathbf{R}}{\partial u_2} \times \frac{\partial \mathbf{R}}{\partial u_3} \right) du_1 du_2 du_3$. Recall that the triple scalar product gives the signed volume of the parallelepiped with the vectors $\frac{\partial \mathbf{R}}{\partial u_1}$, $\frac{\partial \mathbf{R}}{\partial u_2}$ and $\frac{\partial \mathbf{R}}{\partial u_3}$ as sides.

Chapter 6

Appendices

6.1 Appendix A

1. Show that if u and v are pure quaternions, $uv = -\mathbf{u}\cdot\mathbf{v} + \mathbf{u}\times\mathbf{v}$
   Solution:
   $$\begin{aligned}
   uv &= (u_1\mathbf{i} + u_2\mathbf{j} + u_3\mathbf{k})(v_1\mathbf{i} + v_2\mathbf{j} + v_3\mathbf{k}) \\
   &= (u_1\mathbf{i} + u_2\mathbf{j} + u_3\mathbf{k})v_1\mathbf{i} + (u_1\mathbf{i} + u_2\mathbf{j} + u_3\mathbf{k})v_2\mathbf{j} + (u_1\mathbf{i} + u_2\mathbf{j} + u_3\mathbf{k})v_3\mathbf{k} \\
   &= -u_1v_1 - u_2v_1\mathbf{k} + u_3v_1\mathbf{j} + u_1v_2\mathbf{k} - u_2v_2 - u_3v_2\mathbf{i} - u_1v_3\mathbf{j} + u_2v_3\mathbf{i} - u_3v_3 \\
   &= -(u_1v_1 + u_2v_2 + u_3v_3) + (u_2v_3 - u_3v_2)\mathbf{i} + (u_3v_1 - u_1v_3)\mathbf{j} + (u_1v_2 - u_2v_1)\mathbf{k} \\
   &= -\mathbf{u}\cdot\mathbf{v} + \mathbf{u}\times\mathbf{v}
   \end{aligned}$$

3. (a) Let $\mathbf{n}$ denote a unit vector that is perpendicular to a plane P. Thinking of P as a mirror, show that the reflected image of a vector $\mathbf{v}$ in the mirror is given by $\mathbf{v}' = \mathbf{v} - 2(\mathbf{v}\cdot\mathbf{n})\mathbf{n}$
   (b) Show that this can be written in quaternionic form as $v' = nvn$.
   Solution:
   (a) This problem can be expressed in two dimensions which greatly reduces its complexity. We can see that $\mathbf{v}'$ is the sum of the vectors $\mathbf{v}$ and $\mathbf{A}$. We can easily determine the value of $\mathbf{A}$ since it is minus twice the projection of $\mathbf{v}$ on the normal to P, $\mathbf{n}$. The projection of $\mathbf{v}$ on $\mathbf{n}$ is $(\mathbf{v}\cdot\mathbf{n})\mathbf{n}$. Therefore $\mathbf{v}' = \mathbf{v} + \mathbf{A} = \mathbf{v} - 2(\mathbf{v}\cdot\mathbf{n})\mathbf{n}$.
   (b) With information about the dot product of two quaternions from problem 2 we can change the equation into $v' = v + (vn + nv)n$. We can simplify this equation to $v' = v + vnn + nvn$ and since $nn = -1$, $v' = v - v + nvn = nvn$.

5. If z is a complex number, the exponential e^z is defined by the infinite series
   $$e^z = \sum_{n=0}^{\infty} \frac{z^n}{n!}.$$
   Using the same expression to define e^z when z is a quaternion, let $z = \phi u$, where u is a pure quaternion representing a unit vector and ϕ is an angle, and derive $e^{\phi u} = \cos\phi + \sin\phi u$.
   Solution: Begin by writing the infinite series representing $e^{\phi u}$.
   $$\begin{aligned}
   e^{\phi u} &= \sum_{n=0}^{\infty} \frac{z^n}{n!} \\
   &= 1 + u\phi - \frac{\phi^2}{2!} - u\frac{\phi^3}{3!} + \frac{\phi^4}{4!} - \ldots \\
   &= 1 - \frac{\phi^2}{2!} + \frac{\phi^4}{4!} - \frac{\phi^6}{6!} + \ldots + u(\phi - \frac{\phi^3}{3!} + \frac{\phi^5}{5!} - \ldots) \\
   &= \sum_{n=0}^{\infty} \frac{\phi^{2n}}{2n!} + u\sum_{n=0}^{\infty} \frac{\phi^{2n+1}}{(2n+1)!} \\
   &= \cos\phi + \sin\phi u
   \end{aligned}$$
   Note: We could have also applied Euler's identity to solve this problem in a more direct manner.

6.2 Appendix B

No Problems.

6.3 Appendix C

1. As a rule, the angular momentum **L** is not parallel to the angular velocity ω; but ω is directed along the ith principal axis of I, $\mathbf{L} = I_{ii}\omega$ (not summed). Demonstrate this.
   Solution: The inertia tensor can be diagonalized by changing to a body fixed (principal axes) coordinate system, in which case it has the form $\begin{bmatrix} I_{11} & 0 & 0 \\ 0 & I_{22} & 0 \\ 0 & 0 & I_{33} \end{bmatrix}$. Therefore
   $$\begin{bmatrix} L_1 \\ L_2 \\ L_3 \end{bmatrix} = \begin{bmatrix} I_{11} & 0 & 0 \\ 0 & I_{22} & 0 \\ 0 & 0 & I_{33} \end{bmatrix} \begin{bmatrix} \omega_1 \\ \omega_2 \\ \omega_3 \end{bmatrix}, \text{ and } L = \begin{matrix} I_{11}\omega_1 \\ I_{22}\omega_2 \\ I_{33}\omega_3 \end{matrix} \text{ in this system.}$$
   If ω is directed along one of the ith principal axis of I, then $\omega = \omega_i$ and either $L = I_{11}\omega_1$, $L = I_{22}\omega_2$ or $L = I_{33}\omega_3$ since the other two components will be zero. Thus, $\mathbf{L} = I_{ii}\omega$.

3. Prove theorem C.4.
   Solution: Theorem C.4 states

   Theorem 6.3.0.1. *The total kinetic energy of the system equals the sum of the kinetic energies of the individual particles in the CM system, plus the kinetic energy of a single particle of mass M moving with the center of mass:*
   $$\sum_\alpha \frac{1}{2} m_\alpha |\mathbf{v}_\alpha|^2 = \sum_\alpha \frac{1}{2} m_\alpha \left|\mathbf{v}_\alpha - \frac{d\mathbf{R}_{cm}}{dt}\right|^2 + \frac{1}{2} M \left|\frac{d\mathbf{R}_{cm}}{dt}\right|^2$$

   By C.6 we know that the kinetic energy of a single particle is $K_\alpha = \frac{1}{2} m_\alpha |\mathbf{v}_\alpha|^2$. Therefore, the total kinetic energy of the system is $T = \sum_\alpha T_\alpha$. But $|\mathbf{v}_\alpha|^2 = \mathbf{v}_\alpha \cdot \mathbf{v}_\alpha$. So, $\frac{1}{2} m_\alpha |\mathbf{v}_\alpha|^2 = \frac{1}{2} m_\alpha \mathbf{v}_\alpha \cdot \mathbf{v}_\alpha$. Then, we can rewrite $\mathbf{v}_\alpha$, $\mathbf{v}_\alpha = \mathbf{v}_\alpha - \frac{d\mathbf{R}_{cm}}{dt} + \frac{d\mathbf{R}_{cm}}{dt}$.

   $$\frac{1}{2} m_\alpha \mathbf{v}_\alpha \cdot \mathbf{v}_\alpha = \frac{1}{2} m_\alpha \left(\left(\mathbf{v}_\alpha - \frac{d\mathbf{R}_{cm}}{dt}\right) + \frac{d\mathbf{R}_{cm}}{dt}\right) \cdot \left(\left(\mathbf{v}_\alpha - \frac{d\mathbf{R}_{cm}}{dt}\right) + \frac{d\mathbf{R}_{cm}}{dt}\right)$$
   $$= \frac{1}{2} m_\alpha \left(\left(\mathbf{v}_\alpha - \frac{d\mathbf{F}_{cm}}{dt}\right) \cdot \left(\mathbf{v}_\alpha - \frac{d\mathbf{F}_{cm}}{dt}\right) - 2\left(\mathbf{v}_\alpha - \frac{d\mathbf{F}_{cm}}{dt}\right) \cdot \left(\frac{d\mathbf{F}_{cm}}{dt}\right) + \left(\frac{d\mathbf{F}_{cm}}{dt}\right) \cdot \left(\frac{d\mathbf{F}_{cm}}{dt}\right)\right)$$

   Therefore,
   $$\sum_\alpha \frac{1}{2} m_\alpha |\mathbf{v}_\alpha|^2 = \sum_\alpha \left(\frac{1}{2} m_\alpha |\mathbf{v}_\alpha - \frac{d\mathbf{R}_{cm}}{dt}|^2\right) + \sum_\alpha \left(\frac{1}{2} m_\alpha |\frac{d\mathbf{F}_{cm}}{dt}|^2\right) =$$
   $$\frac{1}{2} M \left|\frac{d\mathbf{R}_{cm}}{dt}\right|^2 + \sum_\alpha \frac{1}{2} m_\alpha \left|\mathbf{v}_\alpha - \frac{d\mathbf{R}_{cm}}{dt}\right|^2$$

6.4 Appendix D

1. Verify the formulas (D.49).
   Solution: Formulas (D.49) are

(a) $\nabla_2^2 e^{-i\lambda|\mathbf{R}_2-\mathbf{R}'|} = \left(-\lambda^2 - \frac{2i\lambda}{|\mathbf{R}_2-\mathbf{R}'|}\right) e^{-i\lambda|\mathbf{R}_2-\mathbf{R}'|}$

(b) $2\nabla_1 \cdot \nabla_2 \frac{1}{|\mathbf{R}_2-\mathbf{R}'|} e^{-i\lambda|\mathbf{R}_2-\mathbf{R}'|} = 2i\lambda \frac{\mathbf{R}_1-\mathbf{R}'}{|\mathbf{R}_1-\mathbf{R}'|^3} \cdot \frac{\mathbf{R}_2-\mathbf{R}'}{|\mathbf{R}_2-\mathbf{R}'|} e^{-i\lambda|\mathbf{R}_2-\mathbf{R}'|}$

Using a straightforward modification of vector identities
3.33 $\nabla_2(|\mathbf{R}_2-\mathbf{R}'|^n) = n|\mathbf{R}_n-\mathbf{R}'|^{n-2}(\mathbf{R}_2-\mathbf{R}')$ and 3.28, where $\phi = \phi(\mathbf{R}_2)$, $\nabla_2 \cdot (\phi(\mathbf{R}_2)\mathbf{R}_2) = \phi(\mathbf{R}_2)\nabla_2 \cdot \mathbf{R}_2 + \mathbf{R}_2 \cdot \nabla_2 \phi(\mathbf{R}_2)$ and writing ∇_2^2 as $\nabla_2 \cdot \nabla_2$, we compute

(a) By the chain rule, $\nabla_2^2 e^{-i\lambda|\mathbf{R}_2-\mathbf{R}'|} = \nabla_2 \cdot \nabla_2 e^{-i\lambda|\mathbf{R}_2-\mathbf{R}'|} = \nabla_2 \cdot \left(-i\lambda \frac{\mathbf{R}_2-\mathbf{R}'}{|\mathbf{R}_2-\mathbf{R}'|} e^{-i\lambda|\mathbf{R}_2-\mathbf{R}'|}\right)$.

Then using the modified 3.28, we find (by the chain and product rules),
$$\nabla_2 \cdot \left(-i\lambda \frac{\mathbf{R}_2-\mathbf{R}'}{|\mathbf{R}_2-\mathbf{R}'|} e^{-i\lambda|\mathbf{R}_2-\mathbf{R}'|}\right) =$$
$$-i\lambda \left[|\mathbf{R}_2-\mathbf{R}'|^{-1} e^{-i\lambda|\mathbf{R}_2-\mathbf{R}'|} \nabla_2 \cdot (\mathbf{R}_2-\mathbf{R}') + (\mathbf{R}_2-\mathbf{R}') \cdot \nabla_2 |\mathbf{R}_2-\mathbf{R}'|^{-1} e^{-i\lambda|\mathbf{R}_2-\mathbf{R}'|}\right]$$
$$= -i\lambda \left[3|\mathbf{R}_2-\mathbf{R}'|^{-1} e^{-i\lambda|\mathbf{R}_2-\mathbf{R}'|} + (\mathbf{R}_2-\mathbf{R}') \cdot (-|\mathbf{R}_2-\mathbf{R}'|^{-3}(\mathbf{R}_2-\mathbf{R}')) e^{-i\lambda|\mathbf{R}_2-\mathbf{R}'|} \right.$$
$$\left. +|\mathbf{R}_2\mathbf{R}'|^{-1}(-i\lambda)(\mathbf{R}_2-\mathbf{R}')|\mathbf{R}_2-\mathbf{R}'|^{-1})\right]$$
$$= \left(-\lambda^2 - \frac{2i\lambda}{|\mathbf{R}_2-\mathbf{R}'|}\right) e^{-i\lambda|\mathbf{R}_2-\mathbf{R}'|}, \text{ where we have used } (\mathbf{R}_2-\mathbf{R}') \cdot (\mathbf{R}_2-\mathbf{R}') = |\mathbf{R}_2-\mathbf{R}'|^2.$$

(b) To verify $2\nabla_1 \cdot \nabla_2 \frac{1}{|\mathbf{R}_1-\mathbf{R}'|} e^{-i\lambda|\mathbf{R}_2-\mathbf{R}'|}$, we can begin by looking at the gradient with respect to R_2.
$\nabla_2 \frac{1}{|\mathbf{R}_1-\mathbf{R}'|} e^{-i\lambda|\mathbf{R}_2-\mathbf{R}'|} = \frac{1}{|\mathbf{R}_1-\mathbf{R}'|} \nabla_2 e^{-i\lambda|\mathbf{R}_2-\mathbf{R}'|} = i\lambda \frac{1}{|\mathbf{R}_1-\mathbf{R}'|} \frac{\mathbf{R}_2-\mathbf{R}'}{|\mathbf{R}_2-\mathbf{R}'|} e^{-i\lambda|bR_2-\mathbf{R}'|}$
Next, we can apply the divergence operator.
$$2\nabla_1 \cdot \left(i\lambda \frac{1}{|\mathbf{R}_1-\mathbf{R}'|} \frac{\mathbf{R}_2-\mathbf{R}'}{|\mathbf{R}_2-\mathbf{R}'|} e^{-i\lambda|\mathbf{R}_2-\mathbf{R}'|}\right) = 2\frac{\mathbf{R}_2-\mathbf{R}'}{|\mathbf{R}_2-\mathbf{R}'|} e^{-i\lambda|\mathbf{R}_2-\mathbf{R}'|} i\lambda \nabla_1 \cdot \left(\frac{1}{|\mathbf{R}_1-\mathbf{R}'|}\right)$$
$$= i2\lambda \frac{\mathbf{R}_1-\mathbf{R}'}{|\mathbf{R}_1-\mathbf{R}'|^3} \frac{\mathbf{R}_2-\mathbf{R}'}{|\mathbf{R}_2-\mathbf{R}'|} e^{-i\lambda|\mathbf{R}_2-\mathbf{R}'|}$$
So, we have shown that $2\nabla_1 \cdot \nabla_2 \frac{1}{|\mathbf{R}_1-\mathbf{R}'|} e^{-i\lambda|\mathbf{R}_2-\mathbf{R}'|} = i2\lambda \frac{\mathbf{R}_1-\mathbf{R}'}{|\mathbf{R}_1-\mathbf{R}'|^3} \frac{\mathbf{R}_2-\mathbf{R}'}{|\mathbf{R}_2-\mathbf{R}'|} e^{-i\lambda|\mathbf{R}_2-\mathbf{R}'|}$ which verifies the second part of equation D.49.

3. In many conductors the currents and fields obey an experimental law known as *Ohm's law*: $\mathbf{j} = \sigma \mathbf{E}$, where σ is a constant depending on the conductor, and is called the conductivity. If Ohm's law holds and $\rho = 0$, show that both $\mathbf{E}$ and $\mathbf{B}$ satisfy the *telegrapher's equation*

$$\nabla^2 \begin{pmatrix} \mathbf{E} \\ \mathbf{B} \end{pmatrix} = \frac{\eta\gamma}{k} \frac{\partial^2}{\partial t^2} \begin{pmatrix} \mathbf{E} \\ \mathbf{B} \end{pmatrix} + 4\pi\eta\gamma\sigma \frac{\partial}{\partial t} \begin{pmatrix} \mathbf{E} \\ \mathbf{B} \end{pmatrix}$$

Solution: If $\rho = 0$, then the equation of continuity $\nabla \cdot \mathbf{j} = -i\omega\rho$ becomes $\nabla \cdot \mathbf{j} = 0$, and $\nabla \cdot \mathbf{E} = 0$. Now using Maxwell's equations

$$\nabla \cdot \mathbf{E} = 0$$
$$\nabla \cdot \mathbf{B} = 0$$
$$\nabla \times \mathbf{E} = -\eta \frac{\partial \mathbf{B}}{\partial t}$$
$$\nabla \times \mathbf{B} = 4\pi\gamma\sigma\mathbf{E} + \frac{\gamma}{k}\frac{\partial \mathbf{E}}{\partial t},$$

where we have assumed Ohm's law holds and substituted $\mathbf{j} = \sigma\mathbf{E}$, we take the curl of the second to last equation $\nabla \times \nabla \times \mathbf{E} = -\eta \nabla \times \frac{\partial \mathbf{B}}{\partial t} = -\eta \frac{\partial}{\partial t} \nabla \times \mathbf{B} = -\eta \frac{\partial}{\partial t}\left(4\pi\gamma\sigma\mathbf{E} + \frac{\gamma}{k}\frac{\partial \mathbf{E}}{\partial t}\right) = -\eta(4\pi\gamma\sigma\frac{\partial \mathbf{E}}{\partial t} + \frac{\gamma}{k}\frac{\partial^2 \mathbf{E}}{\partial t^2})$. Because $\nabla \times (\nabla \times \mathbf{E}) = \nabla(\nabla \cdot \mathbf{E}) - \nabla^2 \mathbf{E}$ and $\nabla \cdot \mathbf{E} = 0$, we have

$$\nabla^2 \mathbf{E} = 4\pi\eta\gamma\sigma \frac{\partial \mathbf{E}}{\partial t} + \frac{\eta\gamma}{k}\frac{\partial^2 \mathbf{E}}{\partial t^2}),$$

which we were to prove.
Likewise, taking the curl of the last equation gives us (with the obvious simplification of the left hand side) $-\nabla^2 \mathbf{B} = 4\pi\gamma\nabla \times \mathbf{E} + \frac{\gamma}{k}\frac{\partial}{\partial t}\nabla \times \mathbf{E}$, which, upon substitution of the second to last Maxwell equation, yields

$$\nabla^2 \mathbf{B} = 4\pi\eta\gamma\sigma \frac{\partial \mathbf{B}}{\partial t} + \frac{\eta\gamma}{k}\frac{\partial^2 \mathbf{B}}{\partial t^2},$$

and we're done.

5. If a wire loop is moved through a magnetic induction field $\mathbf{B}(\mathbf{R})$, the conduction electrons "feel" a force $\eta q \mathbf{v} \times \mathbf{B}$, where $\mathbf{v}$ is the velocity of the wire. However, an observer moving with the wire is unaware of any velocity and postulates that the source of this force is an electric field $\mathbf{E}$. Use the flux transport theorem to analyze this situation, and derive the relation (D.28) from Faraday's law.

Solution: The flux transport theorem for magnetic flux can be written as

$$\frac{d\Phi}{dt} = \int\int_{S_t} \left(\frac{\partial \mathbf{B}}{\partial t} + (\nabla \cdot \mathbf{B})\mathbf{v}\right) \cdot d\mathbf{S} + \oint_{C_t} \mathbf{B} \times \mathbf{v} \cdot d\mathbf{R}$$

where $\Phi = \int\int \mathbf{B}(\mathbf{R}, t) \cdot d\mathbf{S}$ is the flux over S, or, by using Stokes' theorem

$$\frac{d\Phi}{dt} = \int\int_{S_t} \left(\frac{\partial \mathbf{B}}{\partial t} + (\nabla \cdot \mathbf{B})\mathbf{v} + \nabla \times (\mathbf{B} \times \mathbf{v})\right) \cdot d\mathbf{S}.$$

Because the divergence of the magnetic field $\mathbf{B}$ is zero by (D. 23), the flux transport theorem becomes

$$\frac{d\Phi}{dt} = \int\int_{S_t} \left(\frac{\partial \mathbf{B}}{\partial t} + \nabla \times (\mathbf{B} \times \mathbf{v})\right) \cdot d\mathbf{S}.$$

Now the term $\nabla \times (\mathbf{B} \times \mathbf{v})$ describes the flux through the surface due to the movement of the surface through the field. Faraday's law states that

$$\frac{d\Phi}{dt} = \int\int_S \frac{\partial \mathbf{B}}{\partial t} \cdot d\mathbf{S} = -\frac{1}{\alpha}\oint_C \mathbf{E} \cdot d\mathbf{R},$$

where $\Phi = \int\int_S \mathbf{B} \cdot d\mathbf{S}$ is the magnetic flux across S and α is a constant. By Stokes' theorem, this can be rewritten as

$$\int\int_S \frac{\partial \mathbf{B}}{\partial t} \cdot d\mathbf{S} = -\frac{1}{\alpha} \int\int_S \nabla \times \mathbf{E} \cdot d\mathbf{S}.$$

We are to derive relation (D. 28), which is that $\alpha = \eta$. From (D. 17) we know that the force felt by a particle of charge q from the magnetic induction vector $\mathbf{B}$ is $F = \eta q \mathbf{v} \times \mathbf{B}$. If the force is due to an electric field, then $\mathbf{E} = \eta q \mathbf{v} \times \mathbf{B}$. Substituting this into Faraday's law, we find

$$\int\int_S \frac{\partial \mathbf{B}}{\partial t} \cdot d\mathbf{S} = -\frac{1}{\alpha}\int\int_S \nabla \times \eta q \mathbf{v} \times \mathbf{B} \cdot d\mathbf{S}.$$

Comparing Faraday's equation to the flux transport theorem, and for unit charge, we find that $-\frac{1}{\alpha}\int\int_S \eta \nabla \times (\mathbf{v} \times \mathbf{B}) \cdot d\mathbf{S} = -\int\int \nabla \times (\mathbf{v} \times \mathbf{B}) \cdot d\mathbf{S}$, so $\alpha = \eta$.

6.5 Appendix E

1. Find the maximum of $f(x, y, z) = -2x^2 - 3y^2 - 4z^2 + 8x + 12y + 24z + 15$.

Solution:
To find a local maximum or minimum we need to find points where the derivative of the function in all directions is zero.

$\frac{df(x,y,z)}{dx} = 0 = -4x + 8$
$\frac{df(x,y,z)}{dy} = 0 = -6y + 12$
$\frac{df(x,y,z)}{dz} = 0 = -8z + 24$

So, the only possible local maximum or minimum occurs at $x = 2$, $y = 2$, $z = 3$. To check if this point is a maximum we can look at the second derivatives which are all negative. This indicates that $(2, 2, 3)$ is indeed a local maximum of $f(x, y, z)$. The value at this point is $f(2, 2, 3) = -2 \cdot (2)^2 - 3 \cdot (2)^2 - 4 \cdot (3)^2 + 8 \cdot 2 + 12 \cdot 2 + 24 \cdot 3 + 15 = 71$.

3. Find the minimum of $f(x, y) = x2 - 4x + y2 - 2y + 5$ subject to the constraint

(a) $x \geq y$

(b) $x \leq y$

Interpret geometrically. [*Hint:* Complete the square for f.]
Solution:

(a) We will use the Kuhn-Tucker equations with one constraint, $\mathbf{grad} f = -\lambda \mathbf{grad} g$.
$\mathbf{grad}(x2 - 4x + y2 - 2y + 5) = -\lambda \mathbf{grad}(x - y)$
This gives us the system of equations,
$2x - 4 = -\lambda$
$2y - 2 = \lambda$
$\lambda \geq 0$

There are two regions we are interested in $\lambda = 0$ and $\lambda > 0$. When $\lambda = 0$,
$2x - 4 = 0$ and $2y - 2 = 0$
which has the solution $x = 2$, $y = 1$ and $f(2, 1) = 0$.
When $\lambda > 0$,
$2x - 4 = -\lambda$
$2y - 2 = \lambda$
$x - y = 0$
which has the solution $\lambda = 1$, $x = 3/2$, $y = 3/2$ and $f(3/2, 3/2) = 1/2$. Thus, the minimum of $f(x, y)$ is at $f(2, 1) = 0$.

(b) We will use the Kuhn-Tucker equations with one constraint, $\mathbf{grad} f = -\lambda \mathbf{grad} g$.
$\mathbf{grad}(x2 - 4x + y2 - 2y + 5) = -\lambda \mathbf{grad}(y - x)$
This gives us the system of equations,
$2x - 4 = \lambda$
$2y - 2 = -\lambda$
$\lambda \geq 0$
There are two regions we are interested in $\lambda = 0$ and $\lambda > 0$. When $\lambda = 0$,
$2x - 4 = 0$, $2y - 2 = 0$ and $y \geq x$
which has no solution.
When $\lambda > 0$,
$2x - 4 = -\lambda$
$2y - 2 = \lambda$
$y = x$
which has the solution $\lambda = 1$, $x = 3/2$, $y = 3/2$ and $f(3/2, 3/2) = 1/2$. Thus, the minimum of $f(x, y)$ is at $f(3/2, 3/2) = 1/2$.

5. Use the Kuhn-Tucker conditions to test which of the following points could be the minimum point for the function $f(x, y)$ whose gradient is given by $(4x + 2y - 10)\mathbf{i} + (2x + 2y - 10)\mathbf{j}$, subject to the constraint $x^2 + y^2 \leq 5$:

   (a) $(1, 2)$
   (b) $(0, 5)$
   (c) $(1, 1)$

   Solution: To begin we need to find the gradient of g. $\mathbf{grad} g(x, y) = \mathbf{grad}(5 - x^2 - y^2) = -2x\mathbf{i} - 2y\mathbf{j}$. Therefore, by Kuhn-Tucker, $4x + 2y - 10 = -2x\lambda$ and $2x + 2y - 10 = -2y\lambda$, $\lambda \geq 0$ and $x^2 + y^2 \leq 5$.

   (a) $(1, 2)$ can be a possible minimum point since it satisfies all of the constraints.
   $4x + 2y - 10 = -2x\lambda$ when $\lambda = 1$, $2x + 2y - 10 = -2y\lambda$ when $\lambda = 1$ and $1^2 + 2^2 \leq 5$.

   (b) $(0, 5)$ cannot be a possible minimum point since it doesn't meet the third constraint.
   $0^2 + 5^2 \not\leq 5$.

   (c) $(1, 1)$ cannot be a possible minimum point since it doesn't satisfy all of the constraints.
   $4x + 2y - 10 = -2x\lambda$ when $\lambda = 2$ but $2x + 2y - 10 = -2y\lambda$ when $\lambda = 3$ which is a problem since λ can not take on two values.

7. Consider the problem of minimizing $f(x, y) = x$, subject to $(x-3)^2 + (y-2)^2 \geq 13$ and $(x-4)^2 + y^2 \leq 16$. Find the three points satisfying the Kuhn-Tucker conditions, and find the minimum.
   Solution:
   $\mathbf{grad}(f) = -\lambda \mathbf{grad}(13 - (x-3)^2 - (y-2)^2) - \mu \mathbf{grad}(16 - (x-4)^2 - y^2)$

   $$\begin{bmatrix} 1 \\ 0 \end{bmatrix} = \begin{bmatrix} 2\lambda(x-3) + 2\mu(x-4) \\ 2\lambda(y-2) + \mu y \end{bmatrix} \quad \text{for } \lambda \geq 0, \mu \geq 0$$

6.5. APPENDIX E

and
$$\begin{bmatrix} 1 \\ 0 \end{bmatrix} = \begin{bmatrix} 2x(\lambda + \mu) - 6\lambda - 8\mu \\ y(2\lambda + \mu) - 4\lambda \end{bmatrix} \text{ for } \lambda \geq 0, \mu \geq 0$$

So, we have six equations.
$2x(\lambda + \mu) - 6\lambda - 8\mu = 1$
$y(2\lambda + \mu) - 4\lambda = 0$
$\lambda \geq 0$
$\mu \geq 0$
$(x-3)^2 + (y-2)^2 \leq 13$
$(x-4)^2 + y^2 \leq 16$

We can see from the first equation that both λ and μ cannot be 0, which gives us three possible points which could be the minimum of f.

Point 1:
$\lambda = 0, \mu > 0$
$2x\mu - 8\mu = 1$
$y\mu = 0$
$(x-4)^2 + y^2 = 16$
We know that μ is not equal to zero so y must be.
$(x-4)^2 = 16$
So, $x = y = 0$.
$f(0, 0) = 0$

Point 2:
$\lambda > 0, \mu = 0$
$2x\lambda - 6\lambda = 1$
$2y - 4 = 0$
$(x-3)^2 + (y-2)^2 = 13$
We can see right away that $y = 2$ so,
$(x-3)^2 = 13$
So, $x = 6.605551275$.
$f(6.6056, 2) = 6.6056$
Note: This point is the maximum value taken on by $f(x, y)$.

Point 3:
$\lambda > 0, \mu > 0$
$2x(\lambda + \mu) - 6\lambda - 8\mu = 1$
$y(2\lambda + \mu) - 4\lambda = 0$
$(x-3)^2 + (y-2)^2 = 13$
$(x-4)^2 + y^2 = 16$
When we try to solve this system of equations we find that it does not have a solution.

So the minimum value of $f(x, y)$ is zero.

9. Prove the statement made in example E.2, that $x = y = 1$ is the only real feasible solution to the Kuhn-Tucker conditions. (*Hint:* The function $x^3 + x - 2$ has a positive slope everywhere, so it can cross the x axis only once.)

Solution: In E.2 we are asked to solve the system of equations,
$$-4x^3 = -\lambda$$
$$-4y = -\lambda$$
$$-16z^3 + 4z = 0$$
$$\lambda > 0$$
$$x + y - 2 = 0$$

We begin by solving the fifth equation for x, $x = 2 - y$, and then substitute this result back into the first equation. Along with some other algebra this gives us,
$$4(2-y)^3 = \lambda$$
$$y = \frac{4 - (2\lambda)^{1/3}}{2}$$

Algebraically we can determine that $y = 1$ and $\lambda = 4$. Substituting y back into the fifth equation we see that $x = 1$. So, with some more work to determine the acceptable values of z, we find that the three possible solutions to E.2 are $(1, 1, 0)$, $(1, 1, 1/2)$ and $(1, 1, -1/2)$ which are the same points determined by the book.

11. How would figure E.7 look if **grad**g and **grad**h were parallel at the point S? What would the Kuhn-Tucker condition for **grad**f become?

Solution: If **grad**g and **grad**h were parallel then **grad**$g = -($**grad**$h)$ and the Kuhn-Tucker condition for **grad**f would be **grad**$f = -\lambda$**grad**g which is the same formula for a problem with only one constraint.